KT-145-412

Fiber Optic COMMUNICATIONS

Lynne D. Green
President
Green Streak Programs
Woodinville, Washington

CRC Press
Boca Raton Ann Arbor London Tokyo

Photo credit: cover photograph #C000-07835-286 Comstock Inc./Michael Stuckey.

Library of Congress Cataloging-in-Publication Data

Green, Lynne D.
Fiber optic communications/Lynne D. Green.
p. cm.
Includes bibliographical references and index.
ISBN 0-8493-4470-0
1. Optical communications. 2. Fiber optics. I. Title.
TK5103.59.G737 1992
621.382′75—dc20 92-36600
CIP

Direct all inquiries to CRC Press, Inc., 2000 Corporate Blvd., N.W., Boca Raton, Florida, 33431.

International Standard Book Number 0-8493-4470

Library of Congress Card Number 92-36600

Printed in the United States of America 1 2 3 4 5 6 7 8 9 0

Printed on acid-free paper

This book is dedicated to those who helped to make it possible. First and foremost is my husband, Kelly Green. Contributions were also made by students and staff at Cogswell College North. The students provided numerous helpful comments while using an early draft as a text. Finally, but by no means least, thanks are due the editors at CRC Press, who brought this book to the light of day.

Table of Contents

PREFACE

This book is an outgrowth of fiber optic design courses given by the author. The focus of the text is the practical design of point–to–point fiber optic links, and is addressed to practicing engineers and students in electrical engineering and electronics technology. The background needed for this material includes engineering mathematics, college physics, and a basic knowledge of analog transistor circuits. A reading list is provided for those desiring to pursue any topic in greater depth.

The text is designed for a senior design–oriented course. It is also suitable for self–study, with numerous examples and a variety of exercises provided. Related topics, such as Maxwell's equations, are included in appendices, together with related exercises.

Applications of fiber optics to digital communications, analog and digitized video, and related areas are discussed. The examples illustrating design equation usage and design techniques draw from these application areas. Design equations are given in the body of the text, together with discussions of their ranges of validity and examples. Typical parameters are used for link components in the examples and the exercises. Component characteristics and design techniques are discussed with a focus towards practical design of fiber optic links.

Components are covered with emphasis on performance within a system. For example, the effect of source spectral width on system bandwidth is discussed in three places: the chapter on modules, the chapter on fibers, and the chapter on design techniques.

A NOTE ON UNITS

SI (MKSA) units are used throughout the book, following standard industry practice. The exception to this rule is for link lengths, which are normally given in either miles or kilometers. A table of physical constants and unit conversions is found in the appendices.

Where industry practice uses standard prefixes, these are also used herein. For example, the wavelength of light is usually given in either nm or μm (e. g. 1.3 μm or 1300 nm), while link lengths are given in km. The reader is encouraged to perform both dimensional and prefix analysis in the exercises as one way of becoming more familiar with the orders of magnitude of typical values encountered in fiber optic link design.

Table of Metric Prefixes

Prefix	Name	Power
T	tera	10^{+12}
G	giga	10^{+9}
M	mega	10^{+6}
k	kilo	10^{+3}
m	milli	10^{-3}
μ	micro	10^{-6}
n	nano	10^{-9}
p	pico	10^{-12}
f	femto	10^{-15}

Chapter 1 Historical Perspective

1.1: Fiber Optic Technology Evolution

Optical communications have been used since early times. Primitive optical communications over free space included the use of signal fires, an early form of digital (one–bit) communication. Polybius, a Greek, later improved on this, using the arrangement shown in Figure 1-1 to spell out words, allowing the transmission of brief messages. Early optical communications were characterized by low information transfer rates and by free space transmission.

Little improvement occurred in optical communications until the 17th century. Isaac Newton [1642–1727], Robert Boyle [1627–1692], and Christiaan Huygens [1629–1695], among others, performed experiments to determine the nature of light throughout the 1600's and early 1700's. Newton discovered dispersion, showing that sunlight was composed of a spectrum of colors. In 1800, William Herschel [1738–1822] discovered infrared energy in sunlight.

Augustine–Jean Fresnel [1788–1827] developed the first detailed mathematical theory of the wave properties of light. He explained for the first time the diffraction effects first observed by Francesco Grimaldi [1618–1663]. Interference effects (e. g. the double–slit experiment) are used today in physics classes to demonstrate the wave properties of light.

The photoelectric effect, observed by Heinrich Hertz [1857–1894] in 1887, showed that light is composed of particles or packets of energy (photons). The cause of the photoelectric effect was explained by Albert

Einstein [1879–1955] in 1905, and he received the Nobel Prize in 1921 for that discovery.

α	β	γ	δ	ε
ζ	η	θ	ι	κ
λ	μ	ν	ξ	ο
π	ρ	σ	τ	υ
φ	χ	ψ	ω	

TORCH PATTERN
Example: μ = Row 3, Col 2

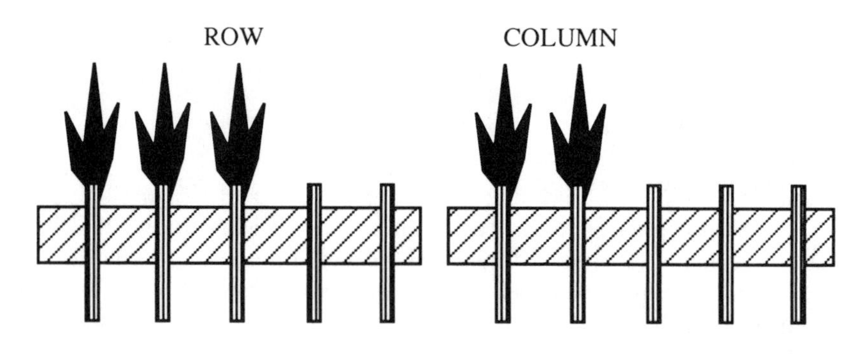

Figure 1-1: Early Greek Optical Communications.

With the development of quantum mechanics in the 1930's, it was realized that light, as well as electrons and other classical particles, have both particle properties (speed, momentum) and wave properties (wavelength, energy). For photons, the wave properties generally dominate, since their rest mass is zero. Quantization (particle) effects become noticeable when photons interact with electrons (as in the photoelectric effect).[1]

Telecommunications began to achieve commercial viability in the 1800's. Samuel Morse [1791–1872] developed a wire–based telegraph in 1838. The telegraph was a digital (short/long pulse) system. Radio operators today still learn the Morse code. The first telephone exchange began operations in 1878.

[1] For classical particles, the particle properties generally dominate, since the speed is much less than that of light. However, wave effects become noticeable when fast particles are diffracted, as in a double slit experiment using electrons.

Maxwell's equations, presented by James Clerk Maxwell [1831–1879], unified the electric and magnetic field concepts of his day. Hertz observed radio waves generated by electric currents in his laboratory. Guglielmo Marconi [1874–1937] made the first demonstration of radio transmission. These led to the development of wireless telegraphy, and later radio communications. Today, free space transmission is used for many transmission systems, such as cellular radio, TV, microwave, and satellite communications.

The concept of guided optical waves is more recent. Total internal reflection of light in a beam of water was demonstrated by John Tyndall [1820–1893] in 1854, but no practical applications for communications were identified at that time. Alexander Graham Bell [1847–1922] invented the photophone in 1880, in which an optical signal was modulated by a membrane which vibrated in response to sound waves in air. Again this was a free space transmission system. The major element in this invention was the movement of a reflective membrane due to audio pressure waves in air.

The first patent on guided optical communications over a transparent medium (glass) was obtained by AT&T in 1934. Light–emitting diode action had been observed in 1923 by O. V. Lossew, and photodetectors even earlier, but these were in use mainly as instrumentation devices. In 1934, there were no transparent materials with sufficiently low attenuation to make the technology feasible. It was not until the 1960's that the proposed systems could be assembled in a laboratory with practical components.

Optical technology took several major strides forward in the 1960's. The gas laser was proposed in 1958 and developed in 1960, and diode laser action was reported in 1962. Meanwhile, a better understanding of loss mechanisms in glass fiber was being developed. Between 1968 and 1970, the attenuation of glass fiber dropped from over 1000 dB/km to less than 20 dB/km, and Corning patented its fiber making process. The combination of diode lasers and low-loss glass fibers allowed fiber optic communications systems to be developed beginning in the early 1970's.

A wide range of new application areas opened up in the 1980's, as the cost of fiber and transmitter / receiver sets decreased dramatically. In addition to telecommunications and local area networks, fiber optic links can now be found in everything from automobiles to compact disc players. Fiber optic links are also used in instrumentation systems, where they provide immunity from EMI noise.

The optical fiber or active optical component can serve as the sensor, providing novel ways to measure chemical and physical properties. Fiber optic sensors are often smaller, lighter, and faster to respond than other sensors. The pH sensor shown in Figure 1-2 uses an optical fiber to transmit light to a reflective medium and to return the reflected light back to the instrumentation system. This sensor can be used in place of litmus paper, and can be used more than once. Many sensors on the market today include autocalibration and digital read–outs, making them suitable for automated measurements.

ure 1-2: A pH Sensor.
Th [illegible] this sensor is in the "eye" of the needle. *Courtesy of Research International, Inc.*

1.2: Comparison to Free Space and Coax Systems

Fiber optic communications offers several significant advantages over the alternatives. Unlike TV and other free space transmission systems, there is no problem of bandwidth allocation or restrictions. In the United States today, for example, the only way to start a new TV station in any major metropolitan area is to buy an existing station. Frequency assignments are allocated by the FCC (Federal Communications Commission) for all communications ranges (TV, radio, microwave, cellular telephones, etc.). The FCC regulations cover carrier frequency assignments, EMI limits, and modulation (e. g. AM/FM and HDTV).

The major advantages of fiber optics over wire systems are the low attenuation and high bandwidth available using glass fiber. Because attenuation is low, longer distances can be spanned between repeaters. Because attenuation is independent of frequency over a wide bandwidth, equalization becomes less important. Figure 1-3 shows the attenuation characteristics of coax and fiber. For high–quality glass fiber, the attenuation curve remains flat beyond 1 GHz.

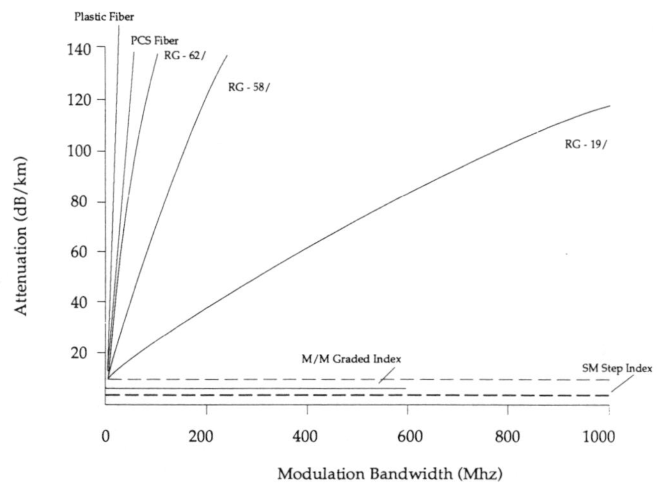

Figure 1-3: Attenuation vs. Frequency.

A comparison of the attenuation of coax and fiber. The fiber has both a lower loss and a flatter frequency dependence. *Courtesy of The Light Brigade Inc.*

For communication of information, there are other advantages as well. With a dielectric (glass or plastic) medium, there are no external electromagnetic fields. Therefore, there is no crosstalk between fibers or EMI (electromagnetic interference) radiated into space. For high bandwidth systems, multiple fibers can be run close together, since there is no crosstalk. For military, banking, and other information systems where data security is important, the lack of EMI provides a system that cannot be tapped with inductive coupling.[1] Dielectric fiber optics prevents emissions from the carrier medium (although it does not prevent board-level emissions).

Because the signal remains completely within the fiber, each user is isolated not only from other users, but also from regulatory agencies. In a fiber optic system, the user is free to select a carrier frequency (or wavelength) and modulation scheme that best fit the system requirements. Regulations and

[1] EMI emission problems can lead to expensive fines. In 1990, for example, one company was fined $4000 for exceeding FCC emission limits for just 2 days; these emissions interfered with a nearby air control tower.

standards are imposed by system requirements rather than by an outside agency.

Fiber optics also presents a number of advantages for industrial and transportation applications. In automated factories, for example, robot systems create severe EMI environments. Robot motors switch large currents through inductive wires, which in turn couple into data network wiring. Data transmission systems can then interpret these signals as false data. Because fiber optics uses a dielectric medium, the EMI signals are not coupled into the fibers, and the data is not corrupted. The dielectric fiber also presents a safety enhancement, since it does not carry current or electrical power. The lack of current means that there is no shock hazard to personnel, and there is no danger of an intermittent open circuit causing a spark. Spark ignition can cause fires and explosions in chemically active environments; these are a prime area for application of fiber optics.

In transportation, the small size and light weight of optical fiber offers a significant advantage. A fiber system may weigh 30 times less than an equivalent copper system. This is also an advantage in city and building ducts, where a 3–in. 900–pair copper cable can be replaced by a single pair of optical fibers with an increase in performance. Figure 1-4 shows the size comparison between this wire cable and an optical fiber. In telecommunications wiring, the copper cable can be replaced with a 144–fiber bundle capable of carrying 245 Gbit/sec. Table 1-1 shows how the wire and fiber cables compare for size, weight, speed, and cost for cables suitable for underground burial. Note that cost varies with year; in 1989 the cost of copper doubled, while the price of fiber decreased.

Cost is the major drawback of fiber optics. For example, connector costs have dropped dramatically, but they continue to be more expensive than twisted pair or coax connectors. Similarly, the cost of fibers, transmitters, and receivers remains somewhat higher than corresponding components for wire-based systems, although the difference is steadily decreasing. For short systems (less than 200 ft.), and for low speed systems (less than 1 Mbit/sec), fiber optics offers little advantage over wire systems.

The cost of a typical telecommunications fiber bundle is seen to be higher than that of copper in Table 1-1. The cost per equivalent telephone circuit-mile, however, is significantly less for a fiber–based system. For large information volumes and/or long distances, the cost advantages of fiber optics are clear. An absolute data rate and distance for the cost break–even point cannot be given, since prices for optical systems are dropping rapidly. Systems costs have dropped as fast as 30% per year, and some active component costs have dropped over 40% in one year.

Although fiber optics offers greatest advantages at long distances and high data rates, the recent introduction of improved plastic fibers has also brought fiber optic networks down to the small office. Plastic fiber can be used to interconnect PCs, printers, and scanners at a cost comparable to that of copper systems, even at relatively short distances. These plastic fiber systems illustrate the tradeoff between cost and performance, being less costly and operating at significantly lower distance and data rates than glass fiber systems.

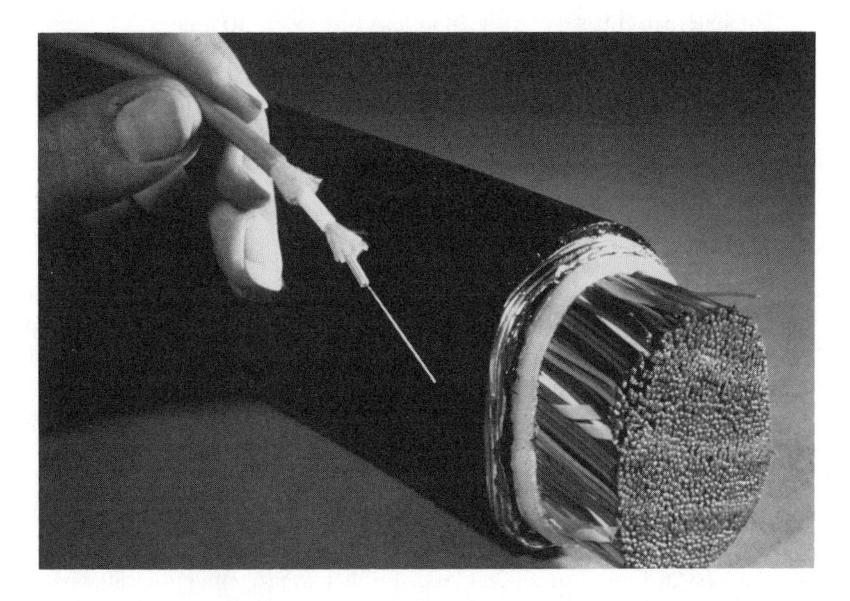

Figure 1-4: Wire Cable and Fiber.

A 900–pair copper cable and a single fiber. The fiber is capable of carrying all of the information of the entire cable. *Photograph courtesy of Corning, Inc.*

A fiber optic system designer may deal with outside agencies when connecting a private system to a larger network. In particular, the telecommunications companies (e. g. AT&T) and other public and private utilities have their own fiber optic network standards. It is possible to lease "dark" (unused) fibers, or to multiplex data onto high-bandwidth lines. For example, the first frame level service was offered to telecommunication users in 1991. Careful attention to the interface standards must be paid if systems integration is to proceed smoothly.

Table 1-1: Comparison of Wire and Fiber Bundles

ITEM	WIRE	FIBER
Count	900 pair (22 G)	144 fibers
Size	2.86 in. diam.	0.5 in. square
Weight	4800 # /1000 ft.	80 # /1000 ft.
Rate / wire	1.54 Mbit/sec	1700 Mbit/sec
Rate / bundle	2.77 Gbit/sec	245 Gbit/sec
Repeater spacing	1.14 miles	30 miles
Cost	$ 8.65 / ft.	$16.56 / ft. (SM)
		$23.22 / ft. (GI)
Cable cost per ckt–mile (1989)	$0.106	$0.023 (SM)

1.3: Technology and Application Trends

Fiber optics evolved as a practical technology in the late 1960's. In this decade, the laser was invented, and light emitting diodes and laser diodes were developed. This decade also saw the introduction of fast integrated circuits, causing system speeds to push the speed limits of wire interconnects.

Major advances in fiber technology were made in the 1960's. Today, fiber attenuation ranges from around 4 dB/km at short infrared wavelengths (0.8 μm) to below 0.2 dB/km at longer wavelengths (1.55 μm). Figure 1-5 shows the attenuation of silica glass fibers; attenuation improved dramatically in the 1970's. Fiber attenuation for high quality fibers is now close to the absolute physical limits for silica glass. Further improvements in attenuation will be achieved through use of other materials, such as fluoride glasses. At the present time, these fibers have not been seen in widespread commercial use.

The availability of high–quality glass fibers provided the first component of a fiber optic system. The 1970's, in turn, was the decade of the active component. Light emitting diodes and laser diodes were designed for higher modulation speeds, better optical efficiency, and increased reliability. The telecommunications industry began installing long–haul links between cities and wide–area networks within major metropolitan areas. One of the biggest advantages for the telecommunications industry was the elimination of repeaters in the "field", moving them into cities and towns. This reduced systems maintenance costs, since repeaters were now easily accessible (e. g. not on mountains or far from roads). Today, telecommunications links span up to 40 miles, so that the majority of repeaters are in town.

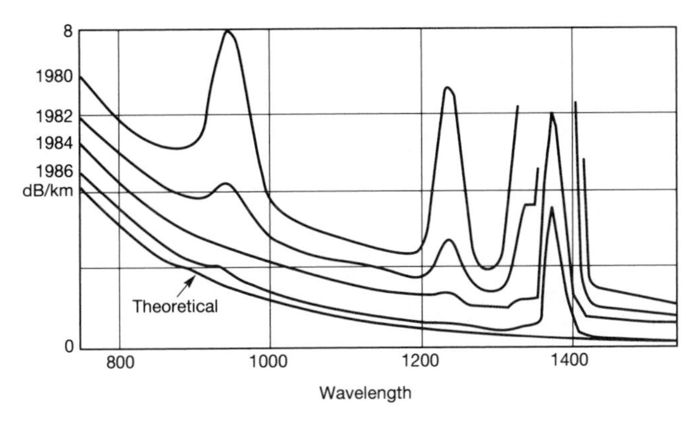

Figure 1-5: History of Attenuation.

Attenuation decreased in the 1970's and 1980's, and has stabilized in the 1990's as fiber fabrication technology has matured. *Courtesy of AMP Inc.*

The 1980's saw the explosion of fiber optics into many application areas. As component costs dropped and physical standards evolved, optical systems became competitive with wire systems for shorter applications. By the middle of the 1980's, fiber optics was competing with satellites and microwave systems. The development of single–mode fiber, and subsequent decrease in cost (Figure 1-6), allowed greatly increased system performance without an increase in cost. Local area networks, spanning up to 2 km, appeared on many university and business campuses. Large factories, such as automobile manufacturing plants, were also "wired" in fiber, as well as other sites where strong EMI problems existed.

The 1980's was the decade of the module. The development of interface standards led to the design of transmitter and receiver modules. The designer no longer designed each module from scratch, saving both time and money. Functional interface modules were available for TTL, ECL, and CMOS connections to the optical system. Functional interface modules, performing numerous support functions (e. g. serial–to–parallel conversion) also appeared on the market.

By the end of the 1980's, cost for fiber and modules decreased to the point that even relatively short links (under 1 km) had become cost effective. The technology for telecommunications, meanwhile, had moved away from the shorter to the longer wavelengths (0.8 to 1.3 and 1.55 μm). Transmitters and receivers had increased performance, with standard modules available up to 12 Gbit/sec. The U.S. telecommunications industry had moved heavily to fiber. Both trans–Atlantic and trans–Pacific fiber cables were in place. Experimental work was underway to wire entire subdivisions with fiber to provide video services as well as telephone service, all over one fiber pair per house.

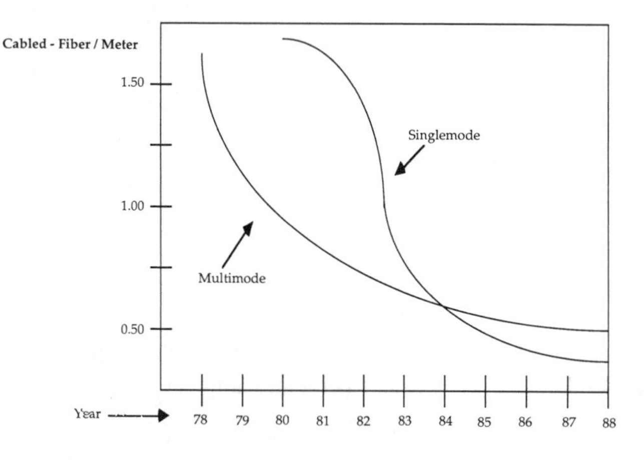

Figure 1-6: History of Fiber Cost.

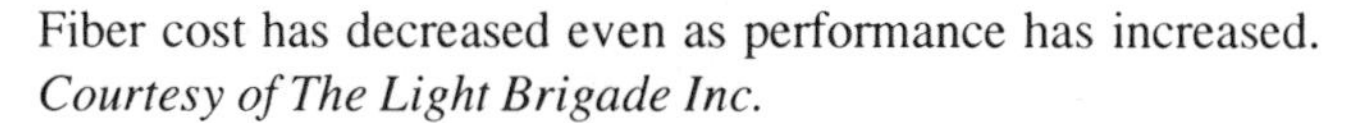

Fiber cost has decreased even as performance has increased.
Courtesy of The Light Brigade Inc.

The 1990's is the decade of the system. The evolution of network standards, standardization of module and link sets, and continuing rapid decrease in cost for standard modules and connectors, are bringing fiber to more applications and end users. Applications will include small office networks in plastic fiber, hybrid fiber and wire networks, and mixed voice / video / data on a network. The trans–Pacific link will be extended across Siberia to meet the trans–Atlantic and European links, completely wiring the world with fiber.

The development of low–cost links has led to use of fiber optics by non–technical end users. As an example, consider the large car dealership with a PC network to track sales and inventory. In the Midwest, a copper LAN might suffer frequent failures due to lightning storms, since lightning grounds itself on conductors. Wired in fiber, this same network will not fail as readily, since lightning will not strike a dielectric. This network, however, will require design, installation, and maintenance by an outside firm. The typical end–user of the 1990's will be non–technical: able to use a PC, but not able to perform systems engineering.

1.4: Market Trends

The technology has moved steadily to longer wavelengths, starting at 800 nm in the early 1970's, moving to 1300 nm in the 1980's, and to 1550 nm today. The shift in wavelength has been accompanied by an increase in systems performance for long–haul links. Cost for the transmitter and receiver modules increases with increasing wavelength. However, the improved transmission distance allows fewer repeaters, resulting in an overall decrease in cost per transmitted bit.

For lower performance networks, the longer wavelength systems do not provide any advantage. These systems will continue to use the 800 nm window as a low–cost solution. At the same time, improved plastic fibers will lead to the use of even shorter wavelengths, in the visible red region (around 600–700 nm) for extremely low–cost systems.

Fiber optics is following many of the same cost and functionality trends as the computer industry. In the computer industry, for example, the cost of a supercomputer system has not decreased significantly since 1960, but the performance has increased spectacularly. At the same time, the basic scientific pocket calculator has not increased in functionality, but the cost has decreased from $1000 to around $15. Much of the increase in performance of high–end systems and decrease in cost of low–end systems has been due to the increasing ability to integrate complete subsystems on a single chip.

Transmitters, receivers, and functional support modules are just reaching full integration as we enter the 1990's. Fully integrated transmitters and receivers emerged from the laboratory in 1989–1990. The first integrated digital fiber optic transmitter–receiver set was also introduced that year. IBM introduced a digital IC set containing internal lasers and photodiodes. The integration of the optical components (LED/laser diodes and photodetector diodes) together with transistor circuits to perform amplification, clocking, encoding/decoding, and other functional communications blocks will allow a significant improvement in cost–performance tradeoffs. Optical integrated circuits and optical components integrated with transistors first moved out of the laboratory into the commercial market around 1990. System designers will see a significant decrease in cost as more vendors introduce module sets to support standard network interfaces.

Cost has decreased for both active and passive fiber optic components. This has been particularly spectacular for single–mode connectors and fibers. As seen in Figure 1-6, the cost of single–mode and graded–index fibers decreased during the 1980's. At this time, there is no indication of fiber cost leveling off in the foreseeable future.

In the early 1980's, a major breakthrough occurred in consumer technology with the introduction of the compact disc audio player. In addition to using fiber optics to isolate digital and sensitive analog circuits, the CD player also used a laser diode to read the optical disk. Today, that same laser, with low cost due to volume production, is used in a variety of small fiber optic networks. As other optical components reach a wider commercial market, similar cost decreases for fiber optic networks may occur.

As performance increases and cost decreases, fiber optics becomes a cost–effective solution to a greater variety of communications applications. In the early 1970's the high cost of active components and fiber restricted fiber optics to those applications where the number of repeaters could be reduced using fiber. In telecommunications, the removal of field–installed repeaters reduced installation and maintenance costs, leading to a lower cost over the life of the system. Because the fiber dominated the systems cost and performance, the telecommunications industry drove the development of improved fiber performance and decreased fiber cost.

In the late 1970's and early 1980's, the decreasing cost of active components and fiber led to the emergence of local area networks (LANs) and metropolitan area networks (MANs). These systems span up to 2 and 10 km, respectively. There was a decreasing use of fiber by the telecommunications industry in the mid–1980's. However, while the volume of fiber used decreased, the volume of active components increased rapidly. The short–haul applications drove the development of lower–cost active components.

Several legal issues influenced the development of fiber optics, and continue to have an impact today. With the breakup of AT&T into regional Bell Operating Companies (RBOCs), there was a separation between cross–continental and local phone service. MANs were developed by both regional RBOCs and by private network owners. One private network issue is the ownership of right–of–way: if the fiber network crosses public utility property (e. g. roads), it comes under government and utility regulation.

As fiber optic systems increase in performance, the range of information that can be transmitted increases. Today, for example, telephone (voice), data, and video information can be transmitted over one fiber using suitable multiplexing schemes. In the United States, however, federal regulations presently prevent the telephone companies from competing with cable TV companies, even though they could share "fiber to the customer". The legal discussions revolve around two major points: ownership (and maintenance responsibility) of the fiber cables and associated electronics, and ownership (content and billing responsibility) of the services carried over the fiber.

One application trend that has only been recently recognized is that many end users do not install or maintain their own fiber optic systems. Just as PC users do not need to understand the electrical functioning of their equipment, so fiber optic communication boards must be "user friendly" to the non–specialist. The user community will require the support of engineering firms in the selection, design, installation, and maintenance of their systems.

To support the commercial end user, two items are required. The first is a self–diagnostic communications system with a support network of engineering support personnel. The other requirement is a set of standards for components and for network management software. Self–diagnostic trouble-shooting is already incorporated in many networks.

1.5: Summary

Optical communications has come a long way since the use of signal fires. Today, fiber optic cable has been installed under the Atlantic and Pacific oceans, across North America and Europe, and will soon be installed around the world. In telecommunications, fiber optics competes with satellites and microwave systems.

Technology has led to the use of three wavelength regions, or "windows": 800, 1300, and 1550 nm. As system operating wavelength increases, both systems performance and cost also increase. The cost for a given performance and the performance for a given cost are both improving with time.

With the decrease in cost and increase in systems performance, fiber optics has become more cost competitive with wire–based systems. This has led to the development of local area networks spanning up to 2 km and metropolitan area networks up to 10 km. Today, fiber optics can be used in links as short as 200 ft. at a cost comparable to that of a wire system.

In the future, fiber optics will find applications in commercial end prod–ucts used by non–technical users. A need presently exists for engineering firms to provide design, installation, and maintenance of these smaller networks.

The next chapter delves into networks to set the stage for the typical operating environment of a fiber optic link. The remaining chapters of this text will cover component performance, system performance analysis, component selection, and systems design. Each topic will be illustrated with examples drawn from a variety of application areas, both those described above and others emerging today. As the cost of fiber optics drops, the range of applications will grow beyond those described in this text, but the systems analysis and design techniques will remain valid.

1.6: Exercises

1. Select one article from the magazines listed in the bibliography, or other sources, which discusses the evolution of fiber optics.
   (a) What technical or legal change is discussed in the article?
   (b) How will this change influence fiber optic applications?

2. Plot cost vs. year for single–mode fiber on semi–log paper. Using a straight–line extrapolation, what is the anticipated cost in 1995? (This has continued to be a good fit to actual cost since about 1982.)
3. Select one article from the above listed magazines which discusses a particular fiber optics application.
   (a) Why is fiber optics the preferred solution for this application problem?
   (b) What are the major advantages *and* disadvantages of fiber optics for this application?

Chapter 2 Network Overview

An information network connects a number of information systems together. The typical network consists of digital point–to–point data links connecting stations (e. g. digital computers). Other networks connect analog information systems, such as sensors and signal processors.

A network's main function is to allow users to share information. This information may be analog or digitized voice or video (e. g. telephone, cable TV), or digital data (e. g. data base information). In general, information is moved across the network in response to a request by a user (computer). The actual movement of data is controlled by the network itself.

A point–to–point link is a single interconnection (line) within a network. In the simplest case, a link is simply a wire between two machines, such as a PC and a printer. In this simple link, the network control is maintained entirely within the PC and printer electronics. In a large network connecting hundreds of computers, there will also be hundreds of links. In this large network, there will be separate network electronics, often on printed circuit boards. At the intermediate level, printed circuit boards can be added to PCs or mini-computers to connect them with a medium sized network. In each of these examples, a point–to–point link connects exactly two "boxes" or stations.

A point–to–point link performs two functions: it formats the data into physical signals, and it moves these signals between two stations. For example, an Ethernet transmitter formats digital data into voltage (logic) levels, and then transmits these voltage signals across a wire to an Ethernet receiver, which converts these signals back into digital data.

The network control electronics (within the sending and receiving stations) provide control signals that indicate when data may be transmitted, when a station is ready to receive data, and when transmission and reception have been completed. In many networks, there is also an ACK, or acknowledge, signal returned to the original transmitting station to verify reception.

When a large amount of data is to be transmitted in a short time, a large system bandwidth is required. This may be achieved by increasing the bandwidth of each point–to–point link, or by increasing the number of links. For example, ASCII data may be transmitted as 8 bits serially on one wire (using time division multiplexing), or 8 separate wires may be used for the 8 separate bits (sometimes referred to as spatial multiplexing). When the data is time division multiplexed, the bandwidth of the single link must be 8 times greater than that of the spatially multiplexed wires.

When multiplexed or system data rates become very high, wire links are unable to move data without high error rates. In this case, fiber optics is used instead of electrical wires. In the telecommunications industry, for example, twisted pair wiring provides sufficient bandwidth for a single telephone conversation, while multiplexing allows 5000 phone calls to be carried on a single fiber pair. Fiber optics is also used for channel extenders, which allow data to move greater distances without repeaters.

Fiber optic links can replace a wire link within a network as long as all electrical interface requirements are met. These interface standards include voltage or current signal levels, timing, and control signals. In general, the physical implementation of the link will be invisible to the computer, system, and end user.

2.1: The OSI Model

The Reference Model of Open Systems Interconnect (OSI), developed by the International Standards Organization, provides a framework for discussing network issues. The seven layers of the model are shown in Figure 2-1. The application and presentation layers deal with user interaction and computer system software. The session layer provides the interface between the user's system and the network, while the transport layer provides the services to format the information into packets. The network layer then routes these packets across the network. The data link layer then breaks the packets into data frames, and manages error correction and acknowledgment signals. The data link layer is also capable of retransmitting any frames that were not correctly received. Finally, the physical layer transmits the formatted information as a bit stream across the actual wire or fiber. It is of interest to note that only the physical link layer contains hardware interface standards.

Many network standards deal with the bottom 3 or 4 layers within the

OSI model. For some standards, particularly proprietary ones, the division between layers does not correspond to the reference model. Network standards developed to use fiber optic links to their best advantage address the bottom 3 layers of the OSI model. The key network issues at the Physical Link layer are the network topology, network access, signal format and the data rate.

	Layer	Function
1	Application	User software interface
2	Presentation	Library software, system routines
3	Session	Communications software to network
4	Transport	System software, data formatting
5	Network	Network software, packet preparation
6	Data Link	Network software, data transmission, error recovery
7	Physical Link	Electrical interface, bit stream transmission

Figure 2-1: Reference Model of Open Systems Interconnect.

Ethernet is a 10 Mbit/sec network standard. It uses a bus topology with CSMA/CD (carrier sense, multiple access with collision detection) control. The bus topology allows any station on the network to use the bus whenever a carrier signal is present. The multiple access with collision detection permits a station to transmit any time the bus is free; if two stations begin transmission at the same time, both stations cease transmitting, waiting some random length of time before trying again.

There are three common logical topologies for networks: bus (e. g. Ethernet), ring (e. g. IBM's Token Ring), and star (e. g. host controller for peripheral units), as illustrated in Figure 2-2. In addition, a network may be connected in one topology logically and another physically. The network shown in Figure 2-3 forms a logical ring (data is transmitted from one station to the next), while the physical layout forms a star connection. Star organization of physical connections allows all wires (or fibers) to be brought to a common patch panel. This makes it easier to add or drop stations or otherwise reconfigure the network.

Within a network, a variety of access methods may be used. A multiple access network, with associated collision detection, allows a station to communicate whenever the network is free. A token passing network requires a station to wait until it has recognized a token (a bit pattern indicating the network is now available to that one station). A star network allows transmission only under the control of one central station.

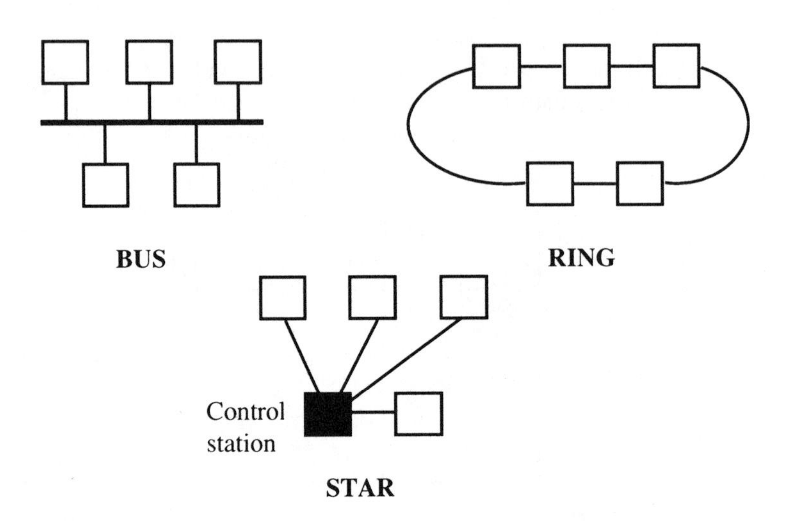

Figure 2-2: Common Network Topologies.

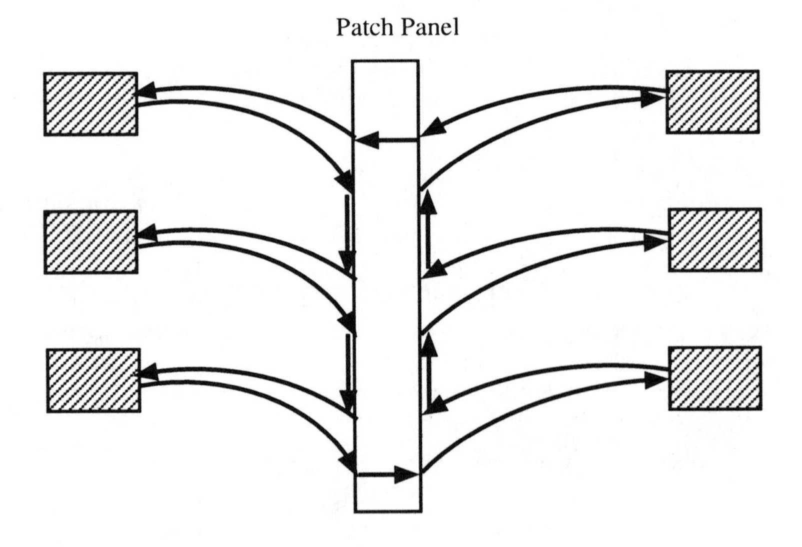

Figure 2-3: A Logical Ring Connected as a Physical Star.

2.2: The Physical Layer

A point–to–point link within a network may be implemented in wire, microwave, fiber optic, or other transmission technologies. The physical implementation is transparent to the software layers. Similarly, the physical layer is insensitive to the logical characteristics of the data.

The lowest level of the OSI model describes the physical transmission medium. Some network standards (e. g. FDDI[1]) specify that a link must be implemented in fiber. Other standards (e. g. Ethernet) do not specify the transmission medium, and either wire or fiber may be used.

Both fiber and wire links may be used within the same network to form what is referred to as a "hybrid" network.[2] For example, a network with widely varying distances between stations may specify wire connections between two nodes closer than 200 ft. and fiber connections for spacings greater than 400 ft. A hybrid network containing a mix of electrical and optical links provides the optimum cost–performance tradeoff for some applications, but at a significant increase in design and maintenance costs.

When fiber optics is to be used in a network, the designer typically specifies transmitter and receiver sets. The sets are matched in their coding scheme, data rate, and optical characteristics. The optical coding used within the set is superimposed upon the digital coding of the network. The second code can be combined with error correction circuits to improve overall network transmission quality.

One key issue that a network designer must consider is that the two ends of any link within the network *must* be technology matched. The electrical interface to the logic circuits (normally TTL/CMOS or ECL levels) is normally the same at the two ends of the link. The two ends of the link must also involve the same transmission media (either wire or fiber). Physical connector styles must also be matched. For example, it is relatively easy to convert between coax and twisted pair connectors. In an optical network, it is more difficult (and expensive) to put jumpers between different connectors.

2.3: Data Coding Techniques

There are many ways of superimposing information on a carrier. In television, analog signals with a 5–MHz bandwidth are used to amplitude modulate a 500–MHz carrier. The high frequency of the carrier allows the signal to

---

[1] Fiber Distributed Data Interface.

[2] The term "hybrid" is also used to refer to a mixture of protocols, topologies, or other technology mixtures.

be transmitted in free space as a broadcast to entire regions. In telephone systems, the analog voice signal is digitized and transmitted by modulating the current or voltage on a wire line. Multiple telephone lines can also be digitally multiplexed and transmitted over an optical link; in this case, the optical carrier frequency is around 2 x 10^{14} Hz. Unlike television systems, each individual telephone connection is distinct, and broadcast (many–party line or public–conference) connections are not used.

Timing is a major consideration in any long–distance communication system. In TV broadcasts, there is a synchronization signal included in the broadcast signal. For telephone networks, time division multiplexing is employed, requiring synchronization of clocking. The recovery of the timing information is called synchronization or clock recovery.

Timing signals in smaller networks, such as internal to a computer, are usually accomplished by incorporating a clock wire separate from the signal wire. In larger networks, a separate clock signal is generally not provided, since a clock signal carries no information. Clock wires are never used in fiber optics, due to the relatively high cost of the individual links and the relatively low cost of incorporating timing information in the data encoding.

Figure 2-4 shows some typical data streams. In the first stream, there is a frequent change in data state between a "1" and a "0" (or HIGH and LOW). In the second case, the data does not change state for several clock cycles. In general, timing can be based on data transition edges provided that these edges occur with sufficient frequency. The purpose of data coding is to ensure a sufficiency of such edges with a minimum bandwidth overhead.

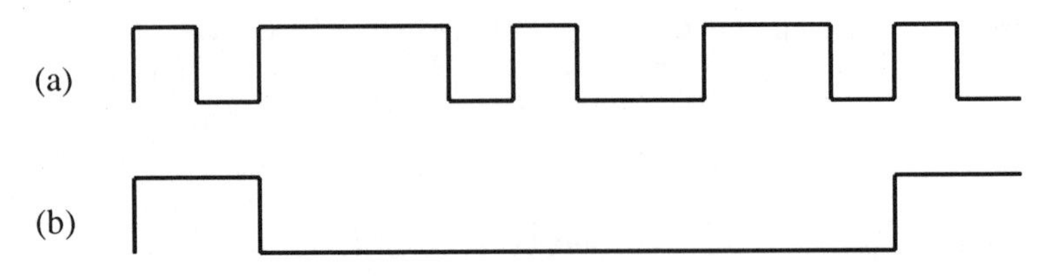

(a) frequent data changes, (b) infrequent data changes.

Figure 2-4: Data Timing.

The minimum frequency for transition edges will depend on the clock recovery circuits used in the receiver. For typical receivers using a phase-locked loop (PLL), timing requires at least one transition every 10 to 15 clock cycles. These transitions can be inherent in the data, as in a parity bit per word, or can be added to the data in the transmitter encoding (and removed in the receiver decoding). Circuits to perform this encoding and decoding are referred to as "codecs". Codecs are required, for example, in telephone networks, where there is no constraint on the length of HIGH or LOW bit sequences; coding of the signal must preserve timing information.

When the encoding scheme imposes a minimum transition frequency (or maximum number of consecutive 1's or 0's), the code is said to be "limited". Most run–length limited codes impose at least one transmission for every 5–10 bits. This encoding may be done by system software prior to passing data to the communications network, on a communications board, or within the transmitter module. In each case, a decoder must appear at the other end of the link, usually at the corresponding communication level (system, board, or module).

There are three codes in common use, and these have become standard for most fiber optic links: (a) NRZ, or Non Return to Zero, which encodes the digital data into an optical data signal without any added transitions; (b) Manchester code, which incorporates a minimum of one transition per bit; and (c) 4B/5B, which encodes 4 bits of data into 5 signal bits incorporating at least one transition. Samples of encoded data are shown in Figure 2-5.

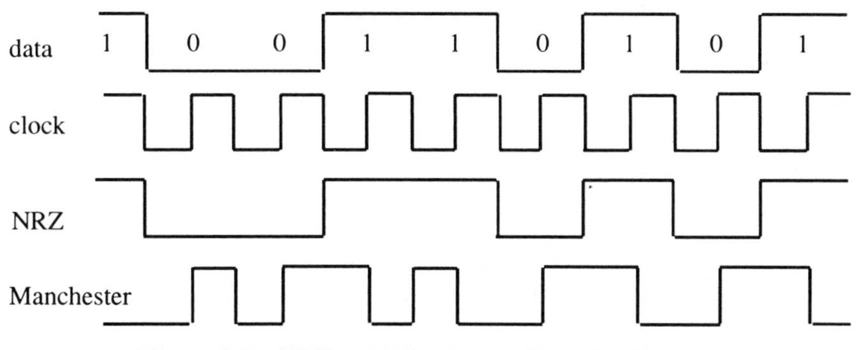

Figure 2-5: NRZ and Manchester Encoded Signals.

The NRZ code is most easily generated from the data. In this case, the digital signal drives a transmitter directly, providing signal voltage and current to a drive circuit. Since the data and the transmitted signal are identical in time, they will have the same frequency components and information bandwidth. Because there are no transitions added in the coding process, the external digital system must provide the necessary run–length[1] limitations in the data format.

[1] Maximum number of consecutive high or low encoded bits.

Manchester code is generated by taking an exclusive OR between the clock and the data (followed by a latch to remove glitches). Since a signal transition may occur on both the leading edge and center of a data bit (both clock edges), the system bandwidth must be double the data bandwidth to accommodate the added transitions.

The 4B/5B code is generated through a table look–up procedure. Four data bits are read in, the table look–up is performed, and 5 signal bits are read out. These bits can then be transmitted with an NRZ transmitter, or can be further encoded if so desired. Because there are 5 signal bits per 4 data bits, the system bandwidth requirement is increased by a factor of 5/4 over the data bandwidth. Data is recovered through a reverse table loop–up at the receiver. The table–lookup procedures are usually supported as a separate IC or incorporated within the transmitter and receiver modules.

Table 2-1 shows the relative efficiency and increase in bandwidth requirements for the most common codes.

Table 2-1: Coding Efficiency and Bandwidth Requirements

Code	Efficiency	B/B_{NRZ}
NRZ	1.00	1.00
Manchester	0.50	2.00
4B/5B	0.80	1.25

EXAMPLE:

A parity system encodes 7 bits of data into an 8–bit word. What are the efficiency and bandwidth requirements for this system?

Efficiency = # signal bits / # data bits = 7 / 8.
Bandwidth = 1/efficiency = 8 / 7 B_{NRZ}.

With the advent of the FDDI standard, costs have decreased and availability increased for chips performing 4B/5B coding. One of the advantages of the 4B/5B code is that some codes are expressly excluded (only 16 of the 32 combinations are required). This means that a power or modulation failure can be distinguished from errors in bit transmission. This enables the system manager to monitor the system bit error rate, used in predictive maintenance.

2.4: The Role of Standards

A system designer must know how components, such as transmitters and receivers, work together. For example, digitized TV signals cannot be displayed directly on an analog TV set. The way in which the information is encoded, bandwidth allocation, and synchronization signals all determine the way in which received information is interpreted.

One example of a communications problem is the composite voice and video TV signal. Vendor A may design a transmitter and receiver set to use direct amplitude modulation for the video portion, with the voice portion FM modulated on a carrier above the video signal. Vendor B may use FM modulation of one carrier frequency for the video and a second carrier frequency for the audio portion. A link designed using a transmitter from vendor A and a receiver from vendor B will not function, even though both ends of the link may have standard NTSC (broadcast) electrical interfaces.

The role of standards is to ensure that parts purchased from different vendors will function in an equivalent fashion. Standards set a minimum level of operational performance and interface characteristics, ensuring interoperability of functional units. Some vendors provide additional functions beyond those required by the standards. These additional functions may or may not have interoperability between vendors.

There are three major standards of interest to fiber optic link designers. These are FDDI (Fiber Distributed Data Interface), SONET (Synchronous Optical NETwork), and B–ISDN (Broadband Integrated Services Digital Network). FDDI is intended for data distribution for large local area networks; SONET is intended for multiplexed voice, video, and data; and ISDN is intended for long–haul public services.

FDDI evolved from a need for higher interconnect speeds between computers located within large area networks. The standard (expected to be finalized in 1992) calls for up to 1000 stations, with a maximum spacing between stations of 2 km; preliminary FDDI standards allowed network installations to begin around 1989. Signaling is at 125 Mbaud, with 100 Mbit/sec of information transmission. FDDI is purely a fiber optic standard. FDDI is packet oriented, and a transmission consisting of several packets may find the packets arriving at the receiving station in any sequence. This makes it almost impossible to transmit voice or video in real time.

A second standard, FDDI–II, is under development. This second standard will address the requirements for integrating analog (voice and video) signals with data signals, with the consequent need for voice and video synchronization. This will allow transmission of real time voice and video signals.

SONET is another purely fiber optic standard. It includes standards for synchronization, allowing voice and video to be transmitted together with data. The basic unit of transmission is the synchronous transport signal at 51.85 Mbit/sec; the other levels are byte–multiplexed sets of these signals. SONET operates at a higher baud rate than FDDI, with a top (OC–48) level modulation of 2488.32 Mbit/sec. SONET is presently the only complete fiber optic standard in widespread use.

B–ISDN evolved from an initial ISDN proposal in the 1980's. In response to the growth of home computing and perceived growth potential in home shopping and interactive educational video programs, ISDN was intended to provide video, voice, and data transmission over optical fibers to individual telephone customers. ISDN baud rates are multiples of the original 64 kbit/sec[1] digitized voice channel. The North American telephone networks now base their systems on multiples of the T–1 (1.544 Mbit/sec) digital carrier. The current B–ISDN proposal calls for 150 Mbaud transmission. Proposals for super–broadband transmission at up to 10 Gbit/sec (10,000 Mbit/sec) have also been made. As with SONET, standards for synchronization of voice and video are under development.

2.5: Network and Data Timing

Timing constraints within a network arise from two limitations. The first, round–trip timing, is constrained by the network software. The second, signal risetime, is constrained by the receiver design as well as system error tolerance.

Round–trip timing is limited whenever an ACK (acknowledgment) signal is returned to the sensing station. If the ACK signal does not arrive within the "time–out" period, it is assumed that the transmission did not arrive, and the entire packet is retransmitted. This time–out period is determined within the network software, and not by any point–to–point link properties.

The round–trip delay can be determined from the formula:

$$T_{\text{trip}} = L_{\text{trip}} / \text{v}$$

where L_{trip} is the round–trip path length the signal must traverse and v is the speed of the signal in the transmission medium.

[1] Audio signals were bandwidth limited to about 3–kHz by the older telephone wire transmission system. The Nyquist criterion sets an 8–k sampling rate for a 4–kHz audio bandwidth. Each sample is taken through an 8–bit analog to digital converter, leading to an overall 64–kbit/sec rate.

EXAMPLE:

An RG/58 coax line carries a signal with a speed of 0.66 c. What is the delay over a path length of 200 ft?

$$T_{trip} = [(200 \text{ ft})(1 \text{ m} / 3.28 \text{ ft})] / [(0.66)(3.0 \times 10^8 \text{ m/sec})] = 308 \text{ nsec}$$

EXAMPLE:

An optical fiber carries carries a signal with a speed of 0.67c. The maximum acceptable round–trip delay is 35 μsec. What is the maximum permissible distance between stations?

$$L_{max} = [(0.67)(3.0 \times 10^8 \text{ m/sec})] [(35 \mu\text{sec})(1 \times 10^{-6} \text{sec} / \mu\text{sec})]$$
$$L_{max} = 7.03 \text{ km}$$
$$\text{maximum spacing} = L_{max} / 2 = 3.5 \text{ km}$$

Data transfer rates within a network are determined by the bit period T_b. The data rate DR is inversely related to the bit period:

$$\text{DR} = 1 / T_b$$

For example, a data transfer rate of 10 Mbit/sec requires a bit period of 100 nsec. Data rates should not be confused with information transfer rates: the overhead added by network software may double the amount of data to be transferred.[1]

Point–to–point link performance may be described in terms of the bit rate, the baud rate, the analog bandwidth, or the rise time (step response). A distinction is made between the data (bit) rate and the baud rate: the baud rate is the number of signal bits/sec, while the data rate is the number of information bits/sec. For example, an FDDI link, using 4B/5B encoding, requires 125 Mbaud to transmit 100 Mbit/sec of information.

The analog bandwidth of a link is the frequency at which the output signal power (i^2) decreases by 3 dB.

Risetimes are used to characterize both analog and digital links. The relationship between risetime and bandwidth (for a system having a single RC time constant) is given as:

$$\tau_r = 0.35 / B$$

---

[1] Addressing, error correction encoding, and other contents are placed in headers and footers, and added to the original user information.

For a digital link, the required risetime is related to the baud rate, and is given by:

$$\tau_r = 0.70 / B$$

In most fiber optic links, the bandwidth is given in MHz or GHz. Link risetimes are typically on the order of a few nanoseconds to a few picoseconds. Note that both transmitter and receiver risetimes must be less than the link risetime. This will be discussed later in the context of link analysis.

EXAMPLE:

A digital data communications system is designed to carry 100 Mbit/sec with NRZ coding. What is the required risetime?

For an NRZ encoded bit stream, the baud rate is the same as the data rate.

$\tau_r = 0.70 / B = 0.70 / 100 \times 10^6$ Hz

$\tau_r = 7.0 \times 10^{-9}$ sec $= 7.0$ nsec.

EXAMPLE:

A digital data communications system is designed to carry 100 Mbit/sec with Manchester coding. What is the required risetime?

For a Manchester encoded bit stream, the baud rate is twice the data rate.

$\tau_r = 0.70 / B = 0.70 / [2 \times 100 \times 10^6$ Hz$]$

$\tau_r = 3.5 \times 10^{-9}$ sec $= 3.5$ nsec.

EXAMPLE:

An analog system has a risetime of 5 nsec. What is the available bandwidth?

$B = 0.35 / \tau_r = 0.35 / 5 \times 10^{-9} = 70$ MHz

The bandwidth of a digital system may be given in a variety of units. The most common are Mbit/sec and Gbit/sec. Systems in which each byte (8 or 10 bits) is transmitted as parallel bit streams may give data rates in Mbyte/sec, in which case the byte width must be known before conversion to bits/sec can be performed. Occasionally, the digital data rate may be given in MHz, with "Hz" representing bits/sec.

At this time, the definition of electrical risetime varies between manufacturers. The most common definitions are for risetime from 10–90% or

20–80% of the signal amplitude. The turn–on time of 0–50% is rarely used; in practice, signals rarely go all the way to zero current or zero voltage. For signals much slower than the available bandwidth, the difference between the two common definitions of risetimes is not significant However, if the design is pushing the bandwidth to its maximum, then this difference must be considered. (Few systems are designed that tightly, as the design becomes excessively sensitive to variations in component performance tolerances.)

Another concern is the frequency content of the signal. For a square wave, for example, the fundamental frequency is $f_0 = 1/(2T_b)$. This square wave will contain signal power at all harmonics (multiples) of this fundamental frequency. However, not all harmonics must be retained to observe the information.

If the presence or absence of square waves is to be observed, then only the fundamental needs to be kept. If the "squareness" of the waveform is of interest, then the first 10 harmonics must be kept. In data communications, the presence of a square wave and its modulation into 1's and 0's is to be observed. This is a case intermediate between observing the presence of modulation and the quality of modulation. The number of harmonics to be retained depends on the nature of the communication system, but is typically only one or two harmonics.[1]

2.6: Measures of Link Performance

There are a number of network performance parameters in addition to the data rate. These issues include signal security and information quality measures. Signal security is of concern primarily in terms of the ability of other stations to receive (tap) information intended for another station; security is primarily a network and system software issue. Security from tapping (e. g. through inductive couplers) can be assured through physical isolation of cables from outside access, and through cryptographic encoding of information.

Link and network information quality is a function of link transmission quality and of error recovery circuit design. Error recovery is accomplished by adding error detection bits to the data stream prior to transmission, and stripping these bits out at the other end of the link. The main effect of error recovery bits is the increase in system bandwidth to accommodate this overhead on top of the user's initial information.

The most direct measure of link data quality is the amount of information received correctly. For an analog link, the quality is measured in

[1] The reader should refer to books on communications theory for further details.

terms of the signal–to–noise ratio (SNR or S/N). The higher the SNR, the higher the link transmission quality. The quality measure for a digital link is the bit error rate (BER). The higher the bit error rate, the lower the link transmission quality. Figure 2-6 shows the relationship between SNR and BER, both for an ideal system and for actual measured receivers. As seen in the figure, SNR is improved by using a single–ended AC–coupled receiver with no hysteresis. A DC–coupled receiver requires about 15 dB more to achieve the same BER as an AC–coupled receiver. Many designers add another 10 dB to compensate for component and circuit performance variations.

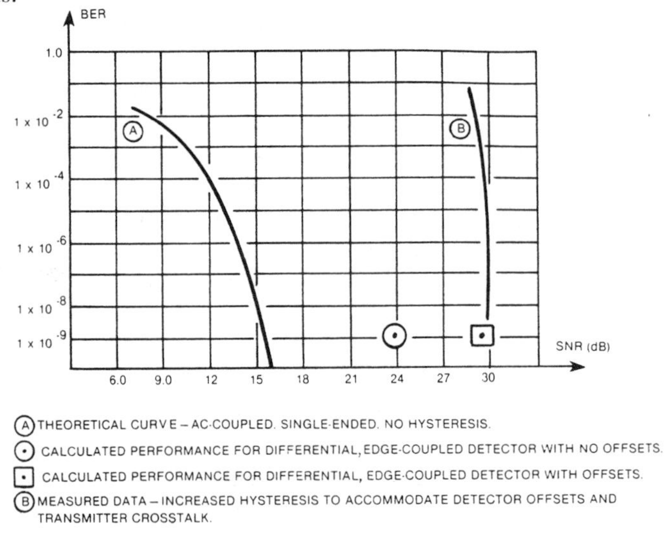

Figure 2-6: BER vs. SNR.

Courtesy of AMP Inc.

The signal–to–noise ratio can be estimated from an "eye pattern" observed on an oscilloscope, such as a sampling oscilloscope. A high–speed eye pattern is shown in Figure 2-7. The eye pattern is named after the opening in the center, which is surrounded by transitions between "1" and "0" data states.

Within the eye pattern, the signal amplitude is the separation of the center of the peak and center of the valley at the maximum eye opening. The noise amplitude is the width of the fuzzy region around the peak. The SNR is the ratio of the square of this signal swing to the square of the noise swing. The timing jitter is the width of the fuzzy region where the 0—>1 and 1—>0 transitions cross. The signal amplitude at the timing crossing is the optimum detection threshold for deciding if the signal is a "1" or a "0". The optimum decision time is midway between the two timing crossings, at the maximum opening of the eye.

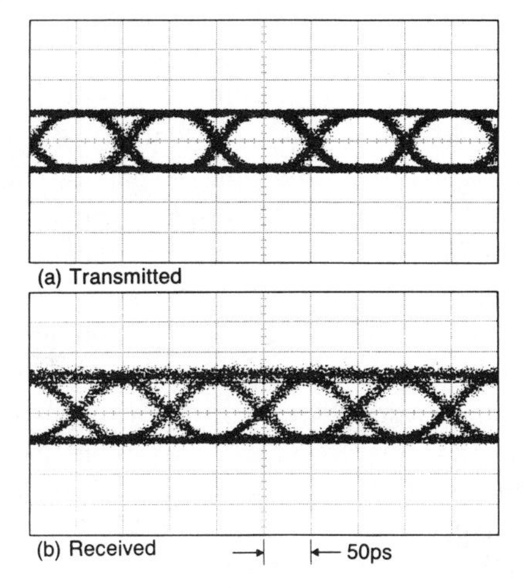

Figure 2-7: Transmitted and Received Eye Patterns.

Reprinted from A. Gnauck, C. Burrus, and D. Ekholm, "A Transimpedance APD Optical Receiver Operating at 10 Gb/sec", *IEEE Photonics Technology Letters*, May 1992. Copyright 1992 by IEEE. Used with permission.

The definition of the signal–to–noise ratio is:

$$\text{SNR} = (I_p / I_n)^2 \qquad \text{relative}$$

$$\text{SNR} = 10 \log [(I_p / I_n)^2] \qquad \text{dB}$$

where I_p is the RMS signal current and I_n is the RMS noise current. (The DC, or average, noise current is normally zero.) The SNR is commonly given in dB.

A loss of information transmission quality may be caused by a number of factors. The two most common problems are (1) a decrease in the amplitude of the received signal, and (2) a loss of timing, often through loss of modulation at the receiver. A loss of signal strength can be determined by measuring received power, while a loss of modulation can be detected by examining the received bit stream.

2.7: Summary

A network is composed of point–to–point link between stations, together with network and system software. A point–to–point link may use twisted pair, coax, microwave, fiber, or other communications technology. Network standards have been developed to ensure interoperability of network components (hardware and software) selected from different vendors.

The three network standards that explicitly call for fiber optic transmission are FDDI, SONET, and B–ISDN. Baud rates for SONET and ISDN networks are based on existing telephone technology, and use digital carriers at multiples of digitized voice signaling rates. Both SONET and ISDN address synchronization issues required for real–time voice and video transmission. FDDI–II, under development, will also address this issue.

Network timing constraints include both round–trip delays and link risetimes. The link risetime limits the maximum baud rate that can be achieved. In addition, coding schemes increase the required baud rate. For example, Manchester encoding requires twice the baud rate as NRZ. The quality of the transmitted data may be measured by the system in terms of the number of errors encountered. At the link level, transmission quality is measured by the BER or SNR.

2.8: Exercises

1. Discuss briefly the various fiber optic standards, and compare them for maximum data rate.
2. A link carries a signal over RG/58 coax. The maximum round–trip time is 250 μsec. What is the maximum station spacing?
3. A ring–connected network contains 25 stations connected by RG/58 coax. The maximum round–trip time is 600 μsec. What is the maximum *average* station spacing?
4. A ring–connected network contains 1000 stations, each with a maximum separation of 2 km. What is the maximum round–trip time if the network uses fiber optic links?
5. An analog system is to carry an information bandwidth of 500 MHz. What is the maximum acceptable link risetime?
6. A digital system is to carry an information bandwidth of 500 Mbit/sec. What is the maximum acceptable link risetime?
7. Compare the risetime requirements of an analog system carrying 10 MHz and a digital system carrying 10 Mbit/sec.
8. A digital system has a risetime of 5 nsec. What is the maximum NRZ data rate?

9. A computer generated video frame containing 1024 x 780 pixels, with 8 color bits per pixel, is stored in a computer file. The file is to be transferred over a network to another computer system.
    (a) How many bits are to be transmitted?
    (b) Assuming no network overhead, how long will it take to transmit this file on an FDDI network? On a high–end SONET network? On a super–broadband ISDN network?
    (c) If the computer image is to be viewed as it is being transmitted (e. g. as part of a sequence of images), the entire file must be transmitted within 1/30 of a second. What is the required network baud rate to achieve this?
10. A link is required to have a BER of 10^{-9}. What is the required SNR?
11. A link has an SNR of 12 dB. What is the best BER that can be achieved?
12. A link has a predicted BER of 10^{-9}. In actual measurements, the link achieves a BER of 10^{-7}. What effects should be investigated as possible causes of the decrease in BER?

Chapter 3 Optics and Materials

3.1: Light

Light has many properties, including both wave and particle characteristics. Wave characteristics include wavelength, frequency, and velocity. Particle characteristics include energy, momentum, and discreteness.

Light may be treated as a wave or a particle, depending on the situation. In general, if the light is propagating through a region or opening much larger than the wavelength, then the light may be treated with classical ray tracing. If the size of the region or opening is smaller than approximately 20 λ, then wave theory (using Maxwell's equations) is required.

The wavelength (λ), frequency (ν), and speed of the light (v) have the relationship:

$$v = \lambda \nu$$

Note that v and ν are different symbols, and should be written carefully to avoid confusion. The frequency ν corresponds to the carrier frequency, as of a TV broadcast system or of an optical carrier.[1] Figure 3-1 shows how λ and ν are defined.

In free space, the speed v is commonly written as **c**:

$$c = 3 \times 10^8 \text{ m / sec} = 3 \times 10^5 \text{ km / sec.}$$

[1] The symbol f will be reserved for the modulation frequency to distinguish it from the carrier frequency.

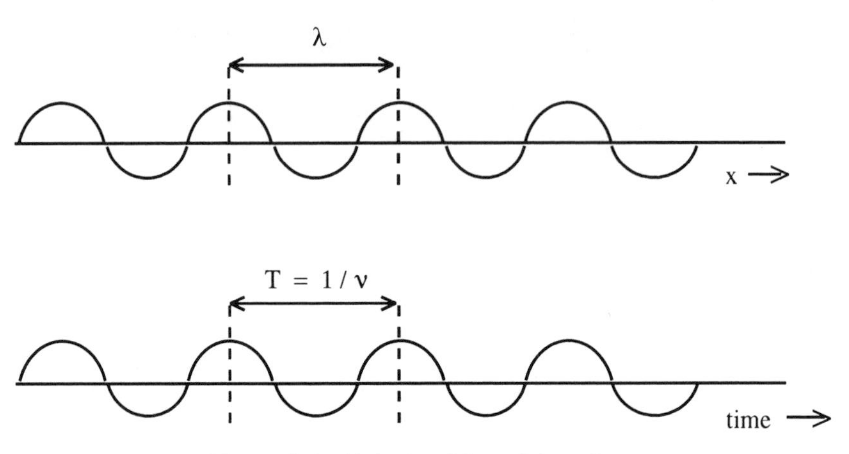

Figure 3-1: Relationships of λ and ν.

In a material, such as glass or water, the speed of light is slower than in free space. In this case, the speed v is related to the speed of light c through the index of refraction (n):

$$v = c / n$$

Table 3-1 shows the index of refraction and relative permittivity for some common materials. Values for other materials can be found in handbooks of chemical and physical properties of materials. There is a simple relationship between n and ε_r for dielectrics:

$$n^2 = \varepsilon_r$$

Table 3-1: Permittivity of Common Materials

Material	ε_r	**n**
Air	1.00	1.00
Diamond		2.42
Crown glass		1.52
Fused quartz		1.46
Water		1.33
Silicon	11.7	
GaAs	13.2	

EXAMPLE:

Determine the index of refraction and speed of light in silicon.

From the table, for silicon $\varepsilon_r = 11.7$

$n^2 = 11.7 \Rightarrow n = 3.42$

$v = c / n = (3 \times 10^8 \text{ m / sec}) / 3.42 = 0.877 \times 10^8 \text{ m / sec}$

3.2: Geometric Optics

When the boundaries of the region under consideration are far away or large compared to the wavelength of the light, geometric or ray–tracing optics can be used. In geometric optics, light is treated as a ray. This ray travels in a straight line in a uniform region of space (region of constant index of refraction). At a boundary between two such regions, the ray is reflected back into the same region and refracted into the second region.

Figure 3-2 shows a ray entering from region I into region II. If the index of refraction is greater in region I than in region II, the ray will be bent away from the normal in region II.[1]

The reflected ray always has the same angle, with respect to the normal, as the entering ray. The energy in the entering ray is divided (unevenly) between the reflected and refracted rays.

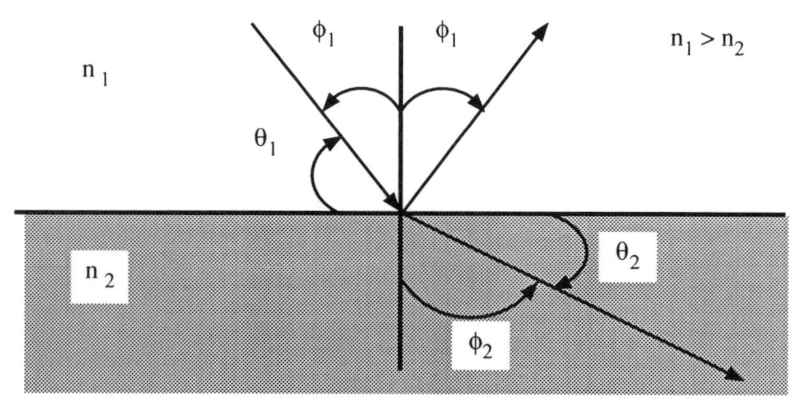

Figure 3-2: Reflection and Refraction at a Surface.

[1] The "normal" is the line perpendicular to the surface at the point of interest. Normals are used in discussing reflection and refraction because the interface may be a curved surface rather than a flat plane.

The refracted ray has an angle determined through Snell's law:

$$n_1 \cos(\theta_1) = n_2 \cos(\theta_2)$$

$$n_1 \sin(\phi_1) = n_2 \sin(\phi_2)$$

As θ_1 approaches 0, θ_2 also approaches 0, as seen in Figure 3-2. When $n_1 < n_2$, $\theta_2 < \theta_1$. For one particular value of θ_1, called the critical angle θ_c, $\theta_2 = 0$. At this angle, the refracted ray ceases to exist. All of the energy in the entering ray is reflected, and the effect is referred to as total internal reflection (since the energy does not cross the boundary). Snell's Law can be used to determine the critical angle:

$$n_1 \cos(\theta_c) = n_2 \cos(0)$$

$$\theta_c = \arccos(n_2 / n_1)$$

In an optical fiber, the index of refraction of the core is always greater than that of the surrounding cladding material. For light entering the core, total internal reflection inside the core will occur for all angles less than the critical angle. One of the major sources of energy loss is removed for such rays, and the optical ray can travel inside the fiber long distances with relatively low attenuation.

Figure 3-3 shows light being coupled from outside a fiber into the core region. Snell's Law can be applied at two interfaces: between the core and cladding, and between the core and air. The external acceptance angle can be determined as follows (noting that the angle to the outside is taken with respect to the normal, and the second form of Snell's Law is used here):

$$n_o \sin(\theta_{o,ext}) = n_1 \sin(\theta_c)$$

$$\theta_{o,ext} = \arcsin[(n_1 / n_o) \sin(\theta_c)]$$

An optical fiber is generally shaped as a long cylinder, with the cladding completely surrounding the core. In this case, the external acceptance angle is the half–angle of a cone through which light is "poured" into the fiber.

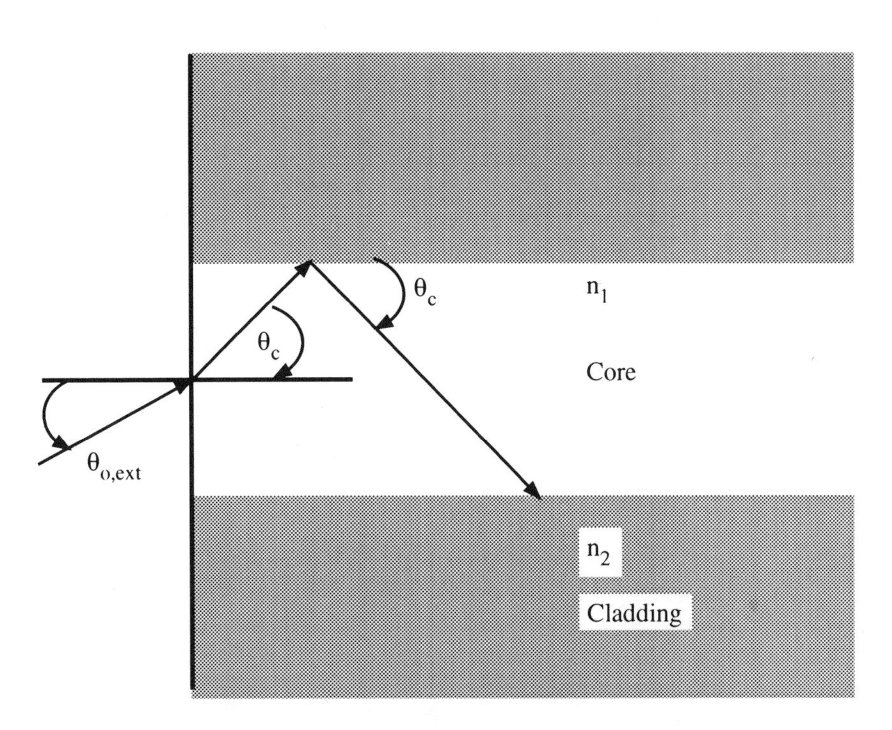

Figure 3-3: The External Acceptance Angle.

EXAMPLE:

A fiber has a core index of 1.446 and a cladding index of 1.430. Light enters from air outside the fiber. Determine the critical angle and external acceptance angle of the fiber.

θ_c = arccos (n_2 / n_1) = arccos (1.430/1.446)
= 8.53^o = 0.142 radians

$\theta_{o,ext}$ = arcsin [$(n_1 / n_o) \sin(\theta_c)$]
= arcsin [(1.446/1.0) sin 8.53^o)]
= 12.4^o = 0.216 radians

3.3: Wave Optics

In the wave theory of light, the electromagnetic properties of the propagating light are explicitly considered. In addition to wavelength and frequency, light also has energy, momentum, and quantization.

Light is composed of photons. Each photon has an associated energy and momentum. For a beam of light, the total energy must be derived from the electric fields, and is dependent on the way in which the (vector) fields add. For light, the total energy E is given by:

$$E = N\,E_p$$

where E_p is the energy of a single photon and N is the number of photons. The power in an optical beam is given by:

$$P = \text{Energy} / \text{Time} = (N / T)\ E_p$$

The momentum of a photon is given by:

$$p = E_p / c = h / \lambda$$

Finally, the energy of a single photon is given by:

$$E_p = h\,c / \lambda = 1.240\ \text{eV–}\mu\text{m} / \lambda$$

EXAMPLE:

An 0.2 mW optical beam is composed of 1300–nm photons. Determine the photon energy and momentum, and the number of photons passing per second.

$$E_p = 1.240\ \text{eV–}\mu\text{m}\ /\ 1.3\ \mu\text{m} = 0.954\ \text{eV}$$
$$p = [(0.954\ \text{eV})\ (1.6 \times 10^{-19}\ \text{J / eV})]\ /\ 3 \times 10^{8}\ \text{m/sec}$$
$$p = 5.09 \times 10^{-28}\ \text{kg-m / sec}$$
$$N / T = P / E_p = 0.2 \times 10^{-3}\ \text{W}\ /\ [(0.954\ \text{eV})\ (1.6 \times 10^{-19}\ \text{J / eV})]$$
$$N / T = 1.31 \times 10^{15}\ \text{photons / sec}$$

Figure 3-4 shows the output spectrum of an optical signal in wavelength space. The optical carrier is characterized by both a center wavelength (λ) and a spectral width (σ_λ). For optical communications the carrier wavelength is usually in the near infrared (between 0.8 and 1.6 µm). The visible spectrum is between 0.7 and 0.4 µm.

The frequency width is related to the spectral width of the light as follows:

$$d\nu = -(c / \lambda^2)\ d\lambda \qquad\qquad d\lambda = -(c / \nu^2)\ d\nu$$

For small spectral widths, as a percentage of the center wavelength or frequency:

$$\Delta\nu = d\nu \qquad\qquad \Delta\lambda = \sigma_\lambda = d\lambda$$

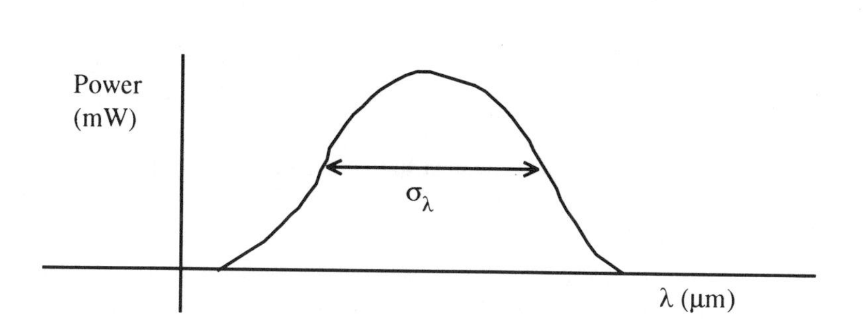

Figure 3-4: Power Spectrum of an Optical Carrier.

EXAMPLE:

A particular LED operating at 820 nm has a spectral width of 45 nm. What is the linewidth in frequency space?

$|d\nu| = (c / \lambda^2)\, d\lambda$
$= [(3 \times 10^8 \text{ m/s}) / (820 \times 10^{-9} \text{ m})^2]\ 45 \times 10^{-9} \text{ m}$
$|d\nu| = 2.0 \times 10^{13}$ Hz
This corresponds to approximately 5.5 % of the center frequency.

The term linewidth may refer to the 3–dB power width in either frequency or wavelength space. It may also be referred to as the FWHM (Full Width at Half Maximum), usually in wavelength space.

The relationships between the electric and magnetic fields comprising the optical signal are related through Maxwell's equations; some of the properties of these relationships are presented in Appendix G. For example, in free space, the electric and magnetic fields are perpendicular to each other. In a metal cavity or waveguide, the boundary conditions require that the electric field go to zero at the cavity edges. In a dielectric cavity or waveguide (e. g. glass fibers), the electric field dies off exponentially outside the core.

Maxwell's equations can be solved exactly for some boundary conditions. In a glass fiber with cylindrical symmetry, the solutions are Bessel functions. The boundary conditions between the core and cladding impose conditions on the characteristics of the solutions. Each solution, or *mode*, can be thought of as corresponding to one ray in geometric optics, although the electric and magnetic fields for that solution exist both inside *and* outside the core region.

For optical communications, there are several characteristics of the solutions that are of interest. First, the number of solutions, or modes, is a finite integer value. Second, for very small core diameters, there is only one solution; these fibers are referred to as *single–mode* fibers. This is in direct contrast to geometric optics, which predicts an unlimited number of modes for all fiber profiles. The modes may be classified by types. Low order modes are those whose energy is contained nearest the center of the core; these are also called bound modes. High order modes are those which contain significant energy (field strength) far from the core.

Another characteristic of the electromagnetic fields is that they extend past the core and into the cladding. The field strength falls off in a roughly exponential fashion in the cladding. The thickness of the cladding must be great enough to contain the fields. Perturbations to the cladding, such as nicks and scratches, will couple power out of the cladding. Bending of the fiber will also cause loss of some of the power propagating in the cladding. For this reason, the highest order modes are also referred to as leaky modes. The low order modes propagate with less attenuation than do the high order modes, since they are less sensitive to the characteristics of the cladding / air interface.

Finally, electromagnetic waves at different wavelengths (or frequencies) propagate independently. Waves interact with other waves only in materials. For radio broadcast systems, for example, each station has an assigned wavelength. All stations broadcast onto free space, and all broadcast signals arrive at an antenna. Separation of the signals occurs in the receiver tuning circuit, so that a radio speaker reproduces only those signals associated with the selected station. Similarly, optical signals at different carrier wavelengths can propagate in a fiber, as long as they can be separated by the photodetector and the receiver circuits.

3.4: Summary

The propagation of light can be described using classical ray or geometric optics when the region of propagation has boundaries large compared to the wavelength. For smaller bounded regions, the wave theory is required.

Ray theory leads to predictions of propagation delays, external acceptance angles, and critical angles for optical fibers. The index of refraction of a material is a measure of the speed of light in that material. Wave theory is used to predict the number of modes and the spatial distribution of the optical power. Wave theory is based on Maxwell's equations.

Multimode fibers carry many hundreds of modes. Single–mode fibers carry only a single mode. Each mode corresponds to one unique solution to Maxwell's equations with the core/cladding boundary conditions. Higher order modes tend to be "leaky", showing higher attenuation and greater sensitivity to the cladding surface and bending than do lower order modes.

Because optical signals do not interact in the fiber, it is possible to have multiple signals on different optical carriers within one fiber. This leads to the use of both electronic and optical multiplexing in high–bandwidth communication systems.

3.5: Exercises

1. Explain why the index of refraction of a material is never less than 1.
2. Complete the entries for Table 3-1.
3. (a) Determine the speed of light in crown glass. (b) How long does it take light to travel 200 ft. in crown glass? (c) To travel 5 miles?
4. An optical fiber has a core diameter of 50 μm, and carries light with a wavelength of 820 nm. Can this fiber be treated using geometric optics?
5. A glass fiber has a core index of 1.456 and a cladding index of 1.448. Determine the critical angle and the external acceptance angle.
6. A glass fiber has a core index of 1.440 and an external acceptance angle of 56^0. Determine the critical angle and the cladding index.
7. An interface has light entering from air at an angle of 45^0 relative to the normal. Light leaves the interface at an angle of 65^0 relative to the interface. What is the index of refraction of this second region?
8. Determine the energy, momentum, and frequency of light having a wavelength of (a) 1300 nm and (b) 850 nm.
9. Compare the energy and momentum of a photon with those of an electron, each having a wavelength of 1300 nm.
10. Compare the momentum of a 1 eV photon to a 1 eV electron. Which is greater? By what ratio?
11. A reference source has a central wavelength of 1300 nm and a linewidth of 1 MHz. What is the spectral width?
12. A laser diode has a spectral width of 2 nm and a wavelength of 1550 nm. (a) What is the linewidth in MHz? (b) What is the spectral width as a percentage of the wavelength? (c) What is the linewidth as a percentage of the optical frequency?

Chapter 4 Transmitters

Transmitters serve to convert electrical inputs into a form suitable for transmission down a line. In fiber optics, the transmitter performs two functions: turning the input signal into a diode drive current, and converting that current into an optical signal. The transmitter contains a light–emitting diode or laser diode, DC bias circuits, and modulation circuits.

Figure 4-1 shows a block diagram of a simplex fiber optic communication link. The transmitter converts electrical signals into modulated optical signals, which are then coupled into the fiber. The receiver converts the optical signal back into an electrical signal, amplifies and demodulates it, and filters it to improve the SNR or BER.

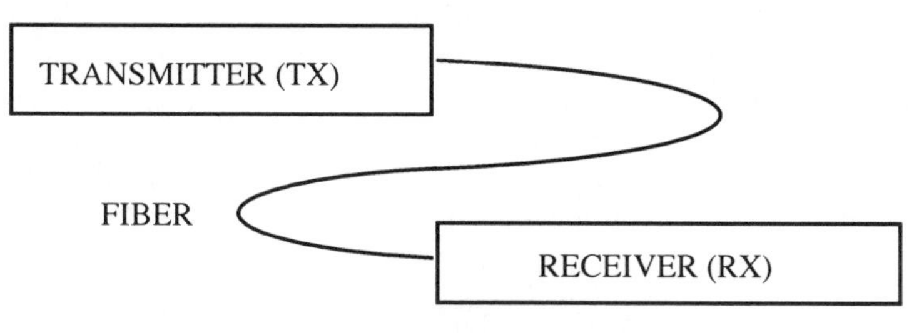

Figure 4-1: System Block Diagram.

For a digital link, the receiver also converts the received signal to a TTL, ECL, or CMOS digital signal level.[1] Figure 4-2 shows a digital link block

[1] TTL: transistor–transistor logic. ECL: emitter coupled logic.
CMOS: complementary MOSFET logic. These are standard digital interfaces.

diagram. In a digital link, the function sections may be incorporated in the transmitter and receiver, reside on separate integrated circuits, or even be built from discrete components.

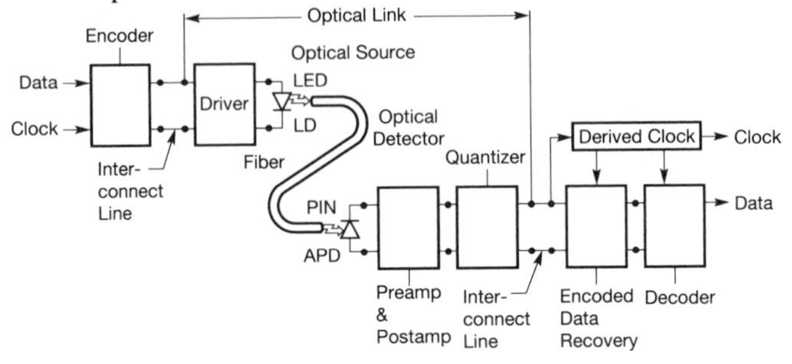

Figure 4-2: Digital Link Block Diagram.

Courtesy of AMP Inc.

4.1: Transmitter Overview

In the early 1970's, transmitters were generally made from discrete devices. In the later 1970's, many transmitters were assembled in hybrid modules containing ICs,[1] discrete resistors and capacitors, and optical source diodes. Encoding the electrical signal (e. g. Manchester coding) was often performed in separate IC modules. A transmitter was assembled by placing the various modules on a printed circuit board.

In the 1980's, transmitter complexity and communication speeds increased. To achieve higher speeds, more of the transmitter function was integrated into VLSI circuits,[2] and the number of board level interconnects was decreased. ASICs were developed by some larger users, particularly in the telecommunications industry. Research efforts were directed at integrating the optical diode and the transistor–level circuits on a common IC substrate without severe loss of performance of either type of device. By 1988, the first integrated receiver and transmitter modules had reached the marketplace.

ASICs also took a second direction in response to the FDDI standard, with a focus on providing ICs to support specific FDDI functions. For example, one IC might perform 4B/5B encoding, while another might provide the parallel–to–serial conversion. Today, FDDI interfaces are small printed circuit boards, but they should appear as single integrated circuits in the near future.

---

[1] IC: integrated circuit. ASIC: applications specific integrated circuit, designed for a specific user or lower volume application.

[2] VLSI: very large scale integration, ICs containing >10,000 gates.

Transmitters continue to be primarily hybrid units, containing both discrete components and integrated circuits in a single package. Figure 4-3 shows a block diagram of a transmitter containing a Peltier cooler and laser power monitor. Figure 4-4 shows a video transmitter, and illustrates the level of complexity that may be found inside the transmitter module. It is also possible to buy discrete optical diodes, and to design a custom transmitter for special applications.

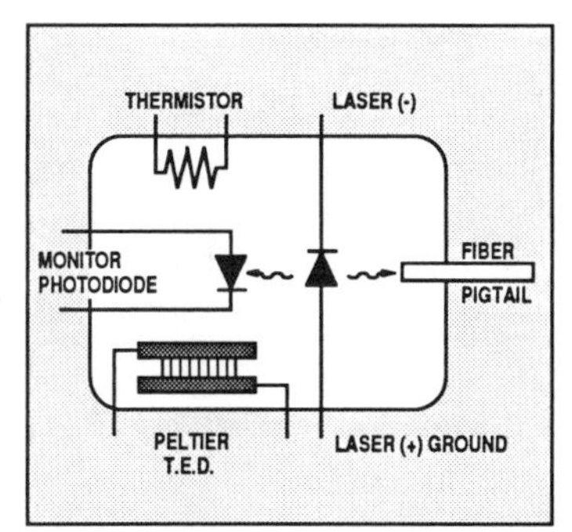

Figure 4-3: Transmitter Package.

Block diagram of transmitter package containing laser, Peltier cooler, and associated circuitry. *Courtesy of BT&D Technologies.*

Figure 4-4: Video Transmitter.

Photograph Courtesy of Math Associates, Inc.

The trend in the 1990's will continue to be the integration of more functionality into ICs for fiber optic links. Chip sets to support video and voice, as well as data, transmission will be developed in response to evolving standards such as ISDN. There will be increasing levels of integration, including multiple lasers and photodiodes on a chip. One such transmitter and receiver set was demonstrated by IBM, and the set included 4 lasers and associated FET components on one IC, and a photodiode, preamplifier, and associated FET logic on the other.

4.2: The Electrical Interface

Most commercial modules use a standard electrical interface. For digital networks, these include TTL, ECL, and CMOS logic levels. Often a transmitter is designed so that the same module can be used for both TTL and CMOS logic. For analog networks, the most common standards are those used in video and television broadcast systems.

Custom input signals can also be handled. This may be accomplished in one of two ways. First, a conversion circuit can accept the input and produce an output compatible with a standard transmitter module. The second alternative is to design and build a custom transmitter module. The first approach is generally used for small volume applications, while the second approach is more suited to large–volume applications.

4.3: Multiplexing Schemes

Many fiber optic links are used to interconnect computers and peripherals. Multiplexing several parallel wire signals into a single serial bit stream allows the transmission of these signals over a single optical fiber. Today's high–speed telecommunications networks, for example, multiplex over 5000 telephone conversations for transmission over a single duplex fiber pair (each fiber carries traffic in one direction). The multiplexers may be contained within the transmitter module or board, or be on a separate board. Corresponding demultiplexing circuits are used at the receiver.

Multiplexing of electronic signals is performed using standard circuits. Among the most common multiplexing schemes are time division multiplexing (TDM), and frequency division multiplexing (FDM). Other electronic multiplexing schemes include pulse modulation schemes such as pulse width or pulse position modulation, and phase modulation.

Time division multiplexing samples a set of inputs sequentially, capturing one input signal in each time bin, and covering all inputs in one complete cycle. For example, a digitizing telephone system samples each voice input at 8–k samples per second (the Nyquist frequency for 4–kHz bandwidth), and converts each sample using an 8–bit A/D converter. The 8–bit samples are serialized, and then the serial streams from several telephones are coupled sequentially onto a single wire or fiber, as shown in Figure 4-5. The North American telephone industry uses the T–carrier digital signaling system. The lowest channel, T–1, carries 24 voice channels on a 1.544 Mbit/sec digital carrier.

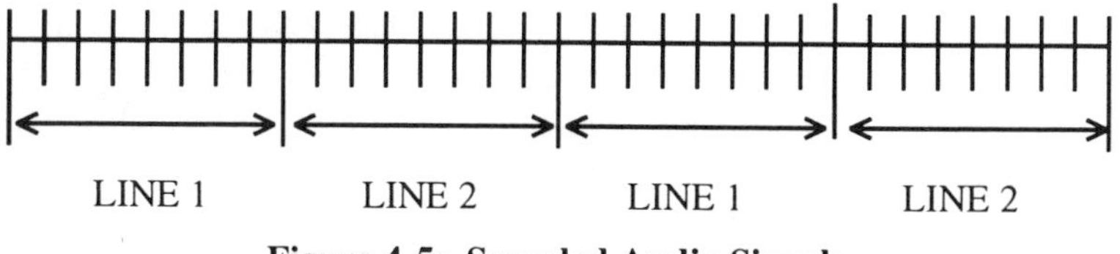

Figure 4-5: Sampled Audio Signals.

Time division multiplexing is particularly suitable for digital signals. Digital multiplexers commonly couple from 2 to 16 input lines to a single (serial) output line. More inputs can be coupled by cascading multiplexers. Because the digital IC industry is relatively mature, time division multiplexing circuits are made from relatively low–cost digital ICs.

Frequency division multiplexing converts baseband amplitude signals into frequency variations around a central carrier frequency. Each input line (channel) has its own separate signal frequency. This is highly desirable for free space broadcast applications. However, FM modulation can not be done using standard digital logic, making it a higher cost multiplexing scheme. FM modulation is used only for low levels of multiplexing in fiber optic systems. The most common use is to shift the audio portion of a TV signal to a carrier frequency above the baseband video signal, allowing both audio and video signals to travel on the same fiber. FM is cost effective when the cost of the analog FM multiplexing is lower than that of the A/D conversion and serializer, and when the SNR of the system is high. At this time, A/D conversion is still quite expensive for video bandwidths and 10–12 bit resolution.

Wavelength division multiplexing (WDM) is a third form of multiplexing, available only in optical systems. It is similar to frequency division multiplexing, except the carriers (frequencies) are multiplexed optically rather than electronically. Each input signal is assigned an optical carrier wavelength, with a corresponding optical frequency. Whereas electronic multiplexing is typically in the range of kHz to MHz, optical carrier frequencies are in the range of 100 GHz or more. There are no electronics available for multiplexing at these very high frequencies.

In WDM, each input's optical signal is generated by a separate optical source. This is similar to TV signals where each TV channel is transmitted from a separate antenna, and TV channels are separated by the tuner at the receiving end. As in TV systems, the wavelengths of the sources must be separated by a sufficient amount to allow them to be separated at the output.

In a fiber optic system, a set of input carriers signals may be independently modulated as long as they can be separated at the receiving end. Because optical fiber is a dielectric, there is no mixing between channels during transmission. For example, digital data can be time division multiplexed on the same fiber with frequency division multiplexed video, if the two sets of information are separated through wavelength division multiplexing. The number of combinations of multiplexing schemes is greater than for typical electronic systems.

4.4: Modules and Discrete Designs

In the early days of fiber optics, all transmitters were made from discrete components: transistors, resistors, capacitors, and light–emitting diodes (LEDs) or laser diodes (LDs). A transmitter might be laid out as a small printed circuit board, or as a more compact thin–film hybrid package. Today, transmitters are usually hybrids, since the smaller package allows shorter wires and therefore higher speed operation. The optical source diode may be inside the same package as the circuitry, or it may be an external component.

Most fiber optic link designs are now done using transmitter modules. This relieves the designer of the need to do extensive high–speed circuit design or to design proper bias networks for optical diodes, since the drive circuits are internal to the transmitter. However, there are limitations to modules. In particular, the electrical interface cannot be changed without changing the transmitter. For example, to change from a TTL to an ECL interface, the module must be changed. This change may involve a new vendor, a new board layout (since the module "footprint" may be different), and perhaps even new bias supply requirements.

The experienced designer may elect to design a transmitter from discrete components and standard ICs, although the higher cost of a custom design will limit this approach to special applications. One such application is a link in which the transmitter end uses TTL and the receiver end ECL digital levels. Since transmitters are normally designed with matched receivers, a mixed digital interface system is difficult to design using standard modules. Other application areas include systems which do not use one of the standard electrical interfaces, and some low–speed and high–speed systems where modules are not available.

4.5: LED and Laser Diodes

Optical sources for fiber optic communications must meet a number of requirements. First, the source's physical size must be comparable to the size of the fiber—on the order of 10–100 μm across. Second, the source must be easily modulated with an electronic signal, and must be capable of high speed modulation. Linearity is also desirable, to prevent harmonics and inter-modulation distortion. Other important requirements include high coupling efficiency to the fiber, high optical output power, small size and low weight, low cost, and high reliability.

Given the size and modulation requirements, the light emitting junction diode is the preferred device. These sources are easily manufactured using standard IC processing, and thousands of diodes can be manufactured on a single wafer. Forward bias conditions are obtained using the same electrical designs as for other diodes, except that higher turn–on voltages are required for optical diodes. The proper selection of semiconductor materials and processing techniques leads to high optical power and efficient coupling of that power to a fiber. Since standard processing techniques are used, low cost and high reliability can be achieved.

The light emitting diode (LED) and laser diode (LD) have electrical characteristics similar to those of other diodes, as shown in Figure 4-6 (a). The optical response to the drive current is very different, however. For a silicon diode, the optical output is zero; this is not a suitable material for an LED. For light emitting materials, the diode structure and material properties will determine whether the device is an LED or a laser diode. The LED has a relatively linear P–I characteristic, as seen in Figure 4-6 (b), while the laser diode shows a strong non–linearity or threshold effect.

Both types of devices may exhibit power saturation (decrease in power at higher current drive levels), and non–linearities in the P–I characteristics. Lasers are also prone to exhibit "kinks", where the power actually decreases with increasing current, as seen in Figure 4-6 (b).

In general, laser diodes have higher modulation bandwidth, greater optical output power, and higher coupling efficiency to the fiber. Light emitting diodes have better linearity, higher reliability, and lower cost. The only exception to the lower cost rule is the compact disc laser. Because this device is used in a high–volume commercial application, its cost is lower than that of other diodes, making it the preferred device for some applications.

Both the LED and LD produce an optical beam which can be coupled to an optical fiber. Each can be modulated with an analog or digital current signal. For analog applications, the LED is preferred because of its better linearity. For high speed digital applications, the LD is preferred because of its higher modulation speeds. Laser diodes can be used for analog applications, but they must be characterized and used only in the linear portion of the P–I characteristic: away from threshold, kinks, and saturation effects.

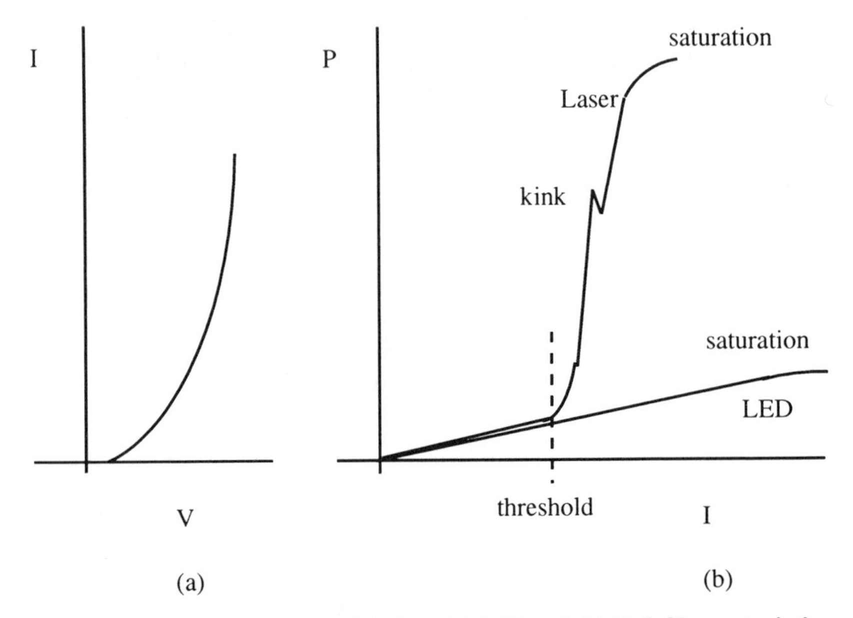

Figure 4-6: LED and Laser Diodes: (a) I–V and (b) P–I Characteristics.

Another key characteristic of the optical output is the wavelength spread over which the power is distributed. The spectral width, σ_λ, is the 3 dB optical power width (usually measured in nm or μm). The spectral width impacts the system bandwidth; a larger spectral width decreases the system bandwidth. In Figure 4-7, the optical power is the area under the curve and the spectral width is the half–power spread. A laser diode always has a smaller spectral width than an LED. The value of the spectral width depends on both device structure and semiconductor material. Typical values are around 40 nm for an LED operating at 850 nm and 80 nm at 1300 nm, and 1 nm for an LD at 850 nm and 3 nm at 1300 nm.

The third key characteristic of junction diodes is the beam distribution, or spatial width, of the output. The more directional the output, the easier it is to couple that power into a fiber. The spatial width is a function of the device structure. A laser diode will have a smaller spatial width than an LED.

The laser characteristics observed in junction laser diodes are significantly different from those of the more familiar HeNe (red) gas laser. The beam spreading in air is far greater for the junction device due to its smaller physical size. The spectral width is also greater for the laser diode. HeNe lasers are used in fiber optics primarily for DC testing of fiber characteristics. Other laser devices can be used, but generally require external modulation, rather than direct modulation of the laser itself. As a result, almost all communication systems now in use use LED or laser diodes, and rarely use other types of lasers. This may change as external modulation techniques evolve and costs decrease.

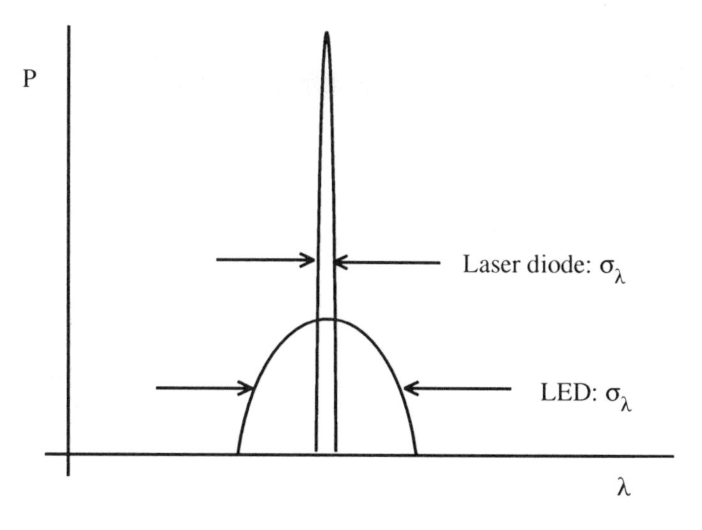

Figure 4-7: LED and Laser Spectral Widths.

4.6: LED Devices

The fundamental LED and laser structure is a junction diode, as shown in Figure 4-8. Electrically, an optical diode acts in the same manner as any other diode. For infrared diodes, the turn–on voltage will typically be 1.0–1.5V. This is significantly higher than the 0.7V typical of a silicon small–signal diode.

Figure 4-8 shows a cross–section of the simplest type of diode. It is referred to as a "homojunction" diode, since both the p and n regions are made from the same semiconductor material. Photons are generated near the junction, and travel in a random direction, as seen by the arrows in Figure 4-8. Electrical contact is made to both the p and n sides of the junction, indicated by the dark contact areas. In a surface view, the upper contact often forms a complete ring, leaving the center open. This is done because metal contacts are generally not transparent. The surface emitting LED, or Burrus diode, emits light at all angles.

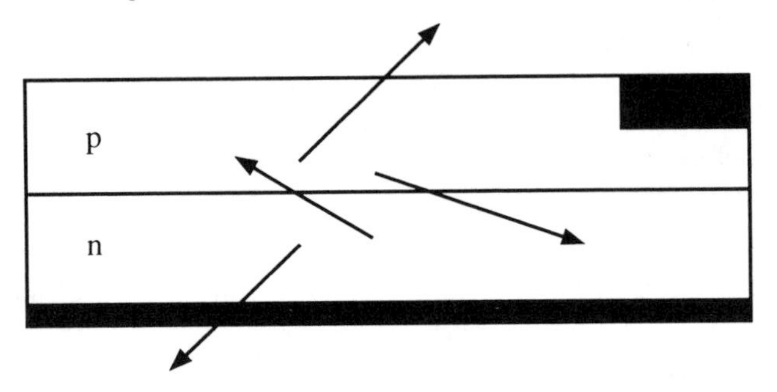

Figure 4-8: A Homojunction Diode.

The second major type of diode is the heterojunction diode, in which the p and n regions are separated by an undoped (intrinsic) semiconductor layer, as shown in Figure 4-9. The regions of wider bandgap (E_{g1}) carry holes (dark circles) and electrons (open circles) to the intrinsic region (E_{g2}). The actual generation of light occurs in the intrinsic region, where the electrons and holes have a high radiative recombination rate. A heterojunction device may emit light from the surface (towards the side of the drawing), or from the edge (into the page).

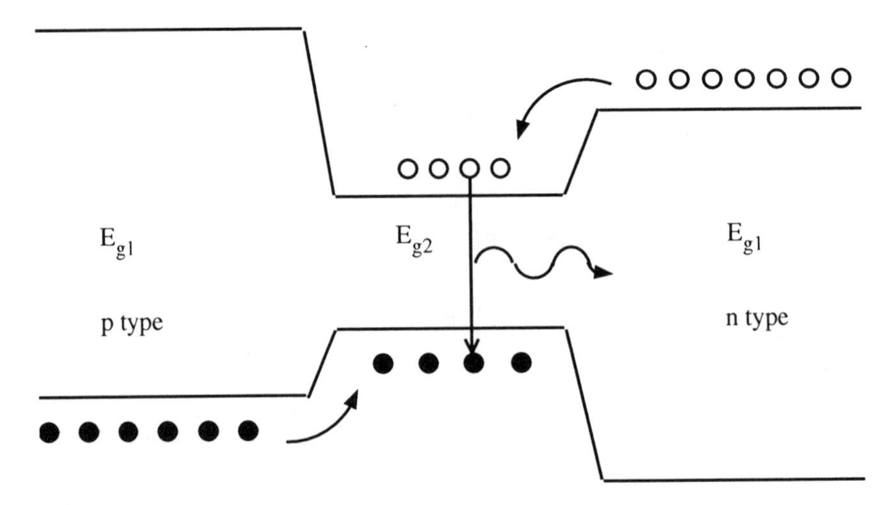

Figure 4-9: A Heterojunction Diode.

The active region of an LED must be made from a direct bandgap semiconductor material. In an indirect bandgap material, such as silicon, the energy involved in electron–hole recombination is released as heat. In a direct bandgap material, such as gallium arsenide, either heat or light can be produced for each recombination. In a high–efficiency device, the majority of recombinations lead to light emission rather than heat. In some compound semiconductor materials, such as $Al_xGa_{(1-x)}As$, the value of the mole fraction x will determine whether the material is a direct or indirect bandgap material. Selected properties of semiconductors are presented in an Appendix.

In an ideal material, one photon of light is emitted for each electron–hole recombination event.[1] The wavelength of the emitted light is determined by the bandgap in the recombination region, according to the formula:

$$E = \frac{h\,c}{\lambda}$$

Light that is emitted by a semiconductor can be absorbed by that same semiconductor material. It is necessary to make the top surface layer of the LED as thin as possible to prevent optical power from being re–absorbed by the LED. This is less of a problem with heterojunction devices, which have layers of wider bandgap materials surrounding the active layer. These wider bandgap materials are transparent to the light emitted from the active layer; also, the region surrounding the active layer is made from an indirect bandgap material, restricting optical emission to the active layer.

Even for heterojunction LEDs, only a small fraction of the photons are emitted from the surface. The amount of emitted light can be increased through the use of reflecting interfaces (similar to mirrors). A reflection occurs whenever light crosses a boundary between two different indexes of refraction.

Reflecting boundaries can be made in several ways. The simplest is the cleaved facet mirror, created by cleaving (breaking) the semiconductor perpendicular to its surface. Because semiconductors are crystalline, the cleaved ends are parallel and flat. Additional chemical polishing can be used to increase the flatness.

Another way to create a reflecting surface is to build a heterojunction diode, either as a two–dimensional "sandwich", or in all three dimensions: surrounding the active area on its sides as well as its top and bottom with wider

---

[1] A hole is the absence of an electron at the site of a chemical bond. A free (conduction) electron is not presently bound into such a bond. In an intrinsic semiconductor, the number of holes and free electrons is relatively low.

bandgap materials. If the wider bandgap material has a lower index of refraction, total internal reflection can occur in the active layer, further increasing the power emitted from the LED. The use of materials of different bandgaps is referred to as bandgap engineering. The use of index differences to provide reflections is referred to as *index guiding*.

A third way to provide a reflective boundary is to confine the current flow in the active region to a narrow stripe, typically by using a metal strip for the top contact, as shown in Figure 4-10. The index of refraction of a semiconductor is dependent on the hole and electron density. When the current is confined, a region of different index of refraction is also created, proving the reflective boundary. This a unique way to create a reflective boundary without also creating a material boundary is referred to as *gain guiding*.

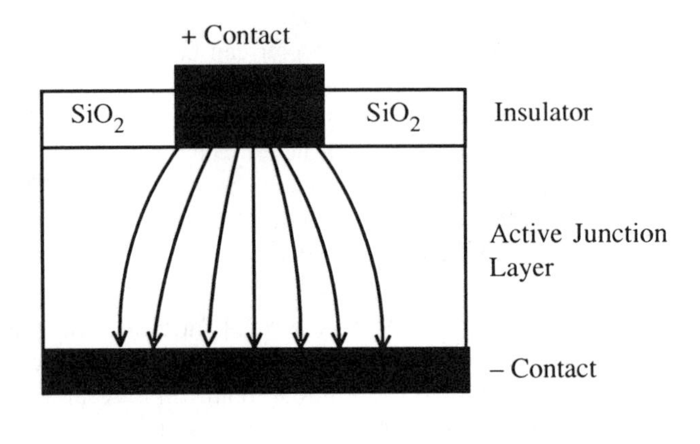

Not to scale.

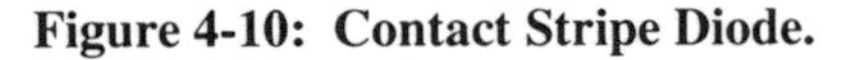

Figure 4-10: Contact Stripe Diode.

When light is reflected back into the active layer from these boundaries, the maximum power is emitted parallel to the active layer's long axis (that is, parallel to the surface). These junctions are used to form edge–emitting LEDs. An edge–emitting LED may produce lower optical power than a surface LED, since the emitting area is small (being the cross–sectional area of the active layer). However, an edge–emitting LED will produce a more directional beam, allowing more efficient coupling of this light into a fiber. Commercial edge-emitting LEDs have a beam width typically 1/3 that of surface LEDs. Edge-emitting LEDs also have a decreased spectral width (typically 30% less than for surface LEDs).

4.7: Laser Devices

A laser diode's structure is very similar to that of an edge–emitting LED. One, two, or even all three of the reflection techniques discussed above may be used within one device. Figure 4-11 shows a device having confinement provided by an ohmic contact stripe, bandgap engineered confining layer, and cleaved end facets. As with edge–emitting LEDs, the light leaves the end faces of the laser.

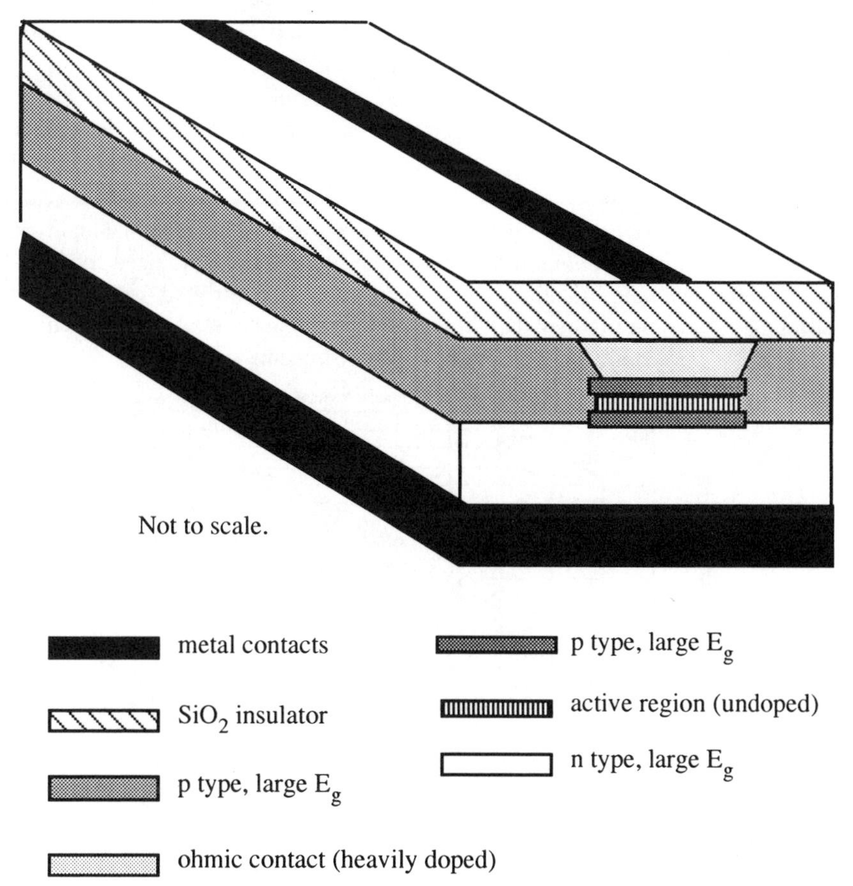

Figure 4-11: Heterojunction Laser Diode Structure.

Laser action is achieved through optical resonance. The active region, with end facets and reflective sides, forms a resonant cavity similar to the Fabry–Perot cavity used for gas lasers. In lasers, photons can be generated with matching frequency, direction, and phase. This is achieved through proper selection of semiconductor materials.

Two types of photon generation occur in a semiconductor. The first type is spontaneous emission, caused by photon emission with random timing, resulting in random phase and direction. Spontaneous emission is characterized by a relatively wide spectral width.

Stimulated emission is caused by photons emitted in response to the presence of another photon. The emitted photon goes in the same direction and with the same frequency and phase of the first photon. In a resonant cavity, those photons traveling along the long axis of the device will collect the largest number of stimulated followers, and will produce a large output beam. In a laser, stimulated emission dominates over spontaneous emission. Laser devices are characterized by highly directional output power distribution and by small spectral widths.

A laser biased below threshold may be used as an optical amplifier. Because the stimulated emission provides the gain, there is no need for high-speed electronic amplification. The gain can be controlled through bias current, as shown in Figure 4-12. These semiconductor optical amplifiers have many of the same optical characteristics as laser transmitter diodes.

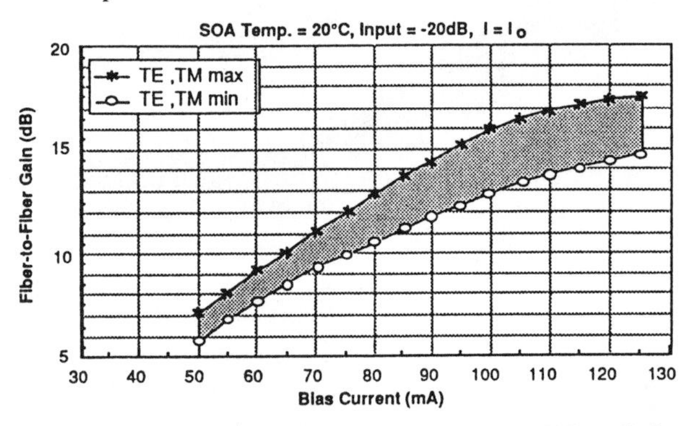

Figure 4-12: Semiconductor Optical Amplifier Gain.

Courtesy of BT&D Technologies.

4.8: Device Biasing

An LED is operated in the forward bias region. The diode may be switched between zero voltage and peak voltage (or zero and peak current) in response to a digital signal. This is the simplest bias scheme, and therefore the least costly to implement.

For higher speed or increased linearity, the LED voltage and current should not drop to zero. The preferred low–end diode voltage is 0.7–1.1 V.

This keeps the diode junction capacitance from fully discharging, while providing a large optical output power swing.

A laser diode is also biased in the forward direction. A DC bias current is used for most transmitters. There is a significant decrease in operating speed if the laser current is allowed to drop below threshold. This is due to the fact that the optical cavity must reach a certain minimum energy density before stimulated emission (laser action) dominates over spontaneous emission (LED action).

The internal gain within the resonant cavity is a function of the DC bias current. If the DC bias is too low, there will not be enough gain to achieve resonance, and the device will act as an edge–emitting LED. A typical laser P–I characteristic shows a strong break point at the threshold current. When a DC bias is applied, time is required to build up the energy density within the cavity. A high–speed laser diode is never driven below its threshold, since this decreases the laser's response speed.

4.9: Environmental Effects

Once the type of optical LED or laser diode has been chosen, additional concerns must be addressed. The concerns include packaging, environmental sensitivity of device characteristics, heat sinking or other thermal regulation, and reliability in the system environment.

The LED or laser package must have a transparent window to transmit light into the air or fiber. A diode may be packaged with a fiber pigtail or with a transparent plastic or glass window. Some companies manufacture packages that have small hemispherical lenses to help focus the light into a fiber.

Packaging must also address thermal coupling for the optical source. Typical LED and laser diodes are driven at 20–80 mA provided from a 5V supply. Assuming a 1.5V drop across the diode, the device consumes as much as 120 mW. The bias resistor may consume almost 300 mW. A complete transmitter module may consume over 1 W. This is a significant power consumption in a small package, and the package must be compatible with heat sink technology. Plastic packages can be used for lower speed transmitters and lower reliability applications. For higher reliability, metal packages with built–in fins for additional heat sinking are preferred.

Figure 4-13 shows the P–I characteristics for a typical laser diode. The operating powers for a logical "0" and logical "1" are also indicated. Note that a logical "0" does not correspond to a physical zero–power level, since the laser is operated above threshold. Note also that the current required to achieve each of the power levels varies dramatically with temperature. If the bias network is designed for a nominal temperature of 20°C, the same current drive will produce two "0" level signals at 70°C.

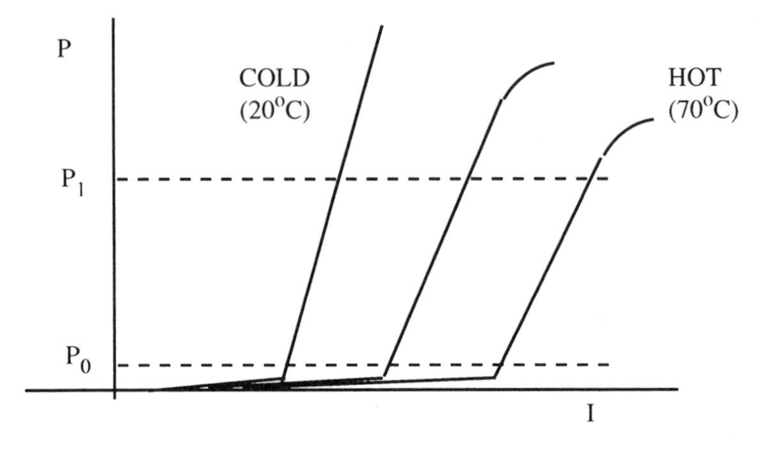

Figure 4-13: Thermal Effects in a Laser Diode.

Lasers require additional thermal control. The DC threshold current is strongly dependent on the temperature. Typical lasers have a threshold current dependence given by:

$$I_{th} = I_z \, e^{(T/T_z)}$$

where I_{th} is the threshold current at temperature T, and I_z and T_z are characteristic of the individual laser diode. Temperatures are in Kelvin.

There are two ways to compensate for the thermal sensitivity of the laser diode characteristics. The most common way is to regulate the diode temperature, usually with a thermoelectric (Peltier) cooler. If the temperature excursions in the environment are not large, it is also possible to use active control of the bias network to keep the "0" level just above threshold.

The active bias network shown in Figure 4-14 takes advantage of the fact that a laser diode emits light from both facets, even though only one facet emits light out of the transmitter module. Back facet monitoring uses a photodiode to measure the optical output from the laser. If the output decreases, feedback circuits increase the DC bias current to the laser. If the output increases, the DC bias current is decreased. If the DC bias current reaches a pre–set limit value, then the laser diode cannot achieve threshold; an indicator is activated, alerting the user that this laser must be replaced.

EXAMPLE:

A laser diode has a threshold current of 20 mA at 25°C and 55 mA at 60°C. Find I_z and T_z.

Taking the ratio of the two threshold currents:

$$\frac{I_1}{I_2} = \frac{e^{\left(\frac{T_1}{T_z}\right)}}{e^{\left(\frac{T_2}{T_z}\right)}} = e^{\left(\frac{T_1 - T_2}{T_z}\right)}$$

Thus,

$$T_z = (T_1 - T_2)\ \ln\left(\frac{I_1}{I_2}\right) = 35\ \ln\frac{55}{20} = 35 \text{ Kelvin}$$

Similarly,

$$I_z = \frac{I_1}{e^{\left(\frac{T_1}{T_z}\right)}} = \frac{20\, x10^{-3}}{e^{\left(\frac{25 + 273}{35}\right)}} = 4.4\ \mu\text{A}$$

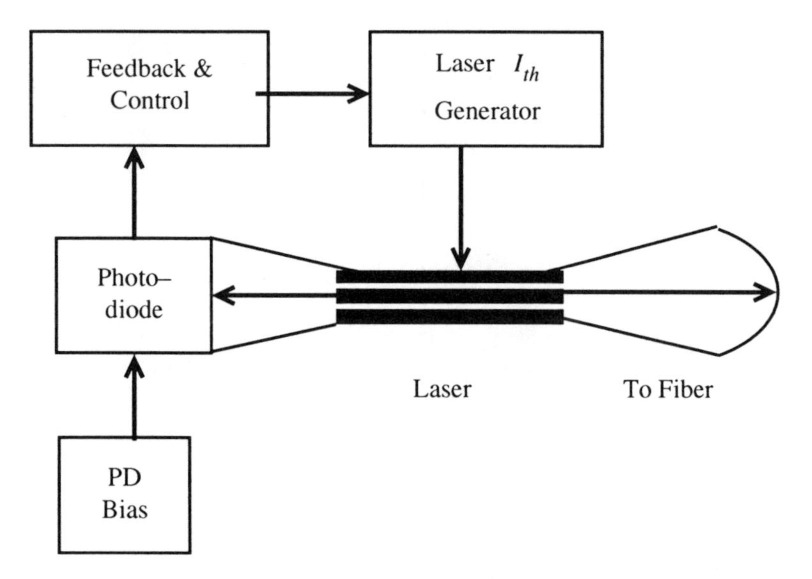

Figure 4-14: Active Bias Block Diagram.

Lasers normally have more than one allowed mode within the active region, similar to that shown in Figure 4-15. The laser spectral spikes fit inside an envelope characteristic of an LED. Usually, one mode will be dominant; however, the particular mode which is dominant may change at random.

The dominant mode can change with temperature, bias current, and modulation depth.[1] Mode changes are referred to as mode hopping. Since the link transfer function is a function of carrier wavelength, a change in the output wavelength introduces a change in the transfer function—which then is interpreted by the system as a change in input. Thus mode hopping is a source of system noise. Mode hopping may be prevented by regulating the device temperature. It can also be prevented by using a DFB (distributed feedback) or a dual–cavity (two–resonance) laser structure. Both of these structures add a second resonance condition to the optical spectrum, eliminating all but one mode in the output spectrum of the laser.

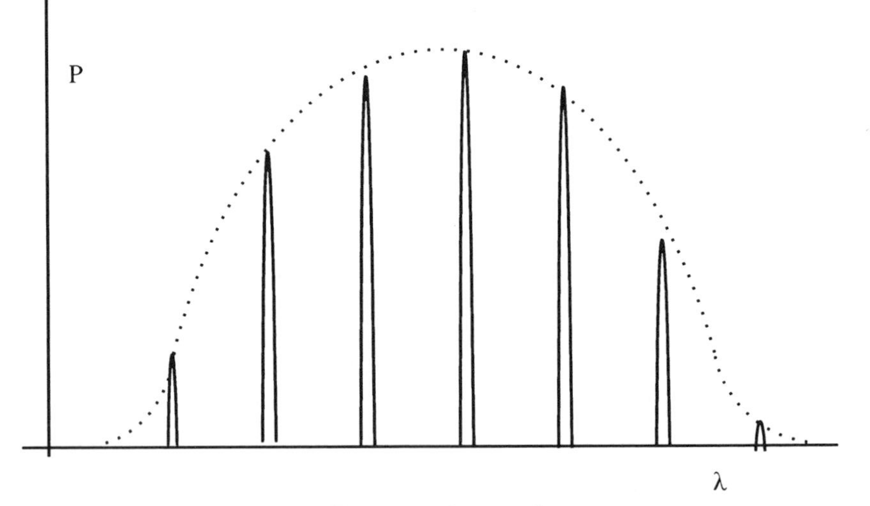

Figure 4-15: Laser Output Spectrum.

The center wavelength of the dominant mode can also be tuned. Tunable lasers are of particular interest for coherent detection systems (in which tuned lasers serve as local oscillators), and for closely spaced wave-length division multiplexing systems. For example, changing the DC control bias in the dual–cavity laser, as shown in Figure 4-16, shifts the resonant

[1] Modulation depth, or modulation index, is the center–to–peak modulation current relative to the bias current (I_{mod} / I_{DC}).

wavelength slightly. Since the output spectrum is dominated by the LED envelope, the tuning range is small. At the present time, commercially available tunable lasers have a range of 40 nm around a 1300–nm center wavelength.

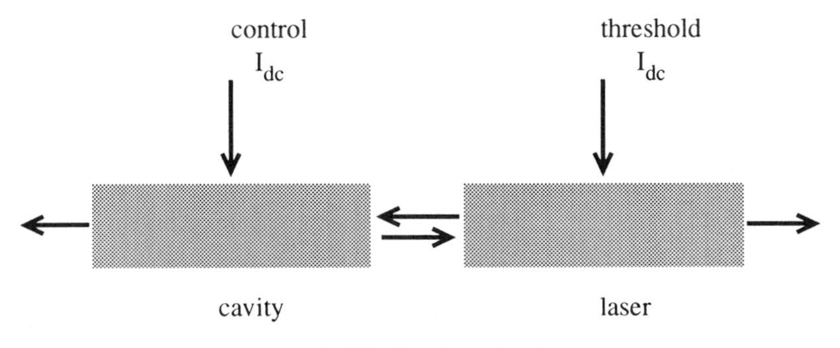

Figure 4-16: Dual–cavity Laser.

4.10: Reliability

Transmitter failures usually occur in the optical source diode. Both LED and laser devices exhibit optical and electrical failures eventually. For both types of devices, a current spike may cause an internal wire to break, resulting in an open circuit. Stress due to overbias may cause the devices to fail as open or as short circuits, depending on the amount of overbias.

Reliability decreases rapidly for increasing temperatures. The mean–time–to–failure can be predicted for any given temperature through the relationship:

$$\text{MTTF} = A\,e^{(E_A/\,kT)}$$

where T is the temperature in Kelvin, E_A is the activation energy of the fault causing failure to occur, and A is a characteristic of the particular device. Typical activation energies for LED and laser diodes are around 0.5–1.2 eV (around 1/2 E_g or so). MTTF decreases with increasing temperature.

Temperature related failures can be minimized by regulating the diode at the lowest possible temperature. Regulation of the temperature below the external (room) temperature using thermoelectric coolers improves reliability for lasers. However, the coolers themselves will also fail. The overall transmitter reliability will depend on the reliability of the cooler, the transmitter electronics, and the laser.

Thermal regulation is usually used with laser diodes, both to reduce mode hopping and to improve reliability. Thermal regulation is generally not required with LEDs except for precision analog applications. In these applications, the thermal changes cause changes in the diode optical output (at constant current), which are interpreted by the system as changes in the input.

MTTF decreases with increasing current, with:

$$\text{MTTF} = MTTF_o \, (I_o / I)^n$$

where I_o and $MTTF_o$ are known at one current, and MTTF is predicted at the new current I. For typical optical diodes, the exponent n is between 1.5 and 2. Operating MTTF at room temperature can be as high as 6 years for laser diodes and 25 years for LEDs. MTTF is significantly shorter at higher operating temperatures and currents.

Because of the decrease in reliability with increasing current, both LEDs and laser diodes are run at the lowest current for which the link will meet specifications. Within a transmitter module, the bias network is designed for the maximum performance specified. Reliability can be increased by decreasing the temperature, as long as module specifications are not exceeded.[1] Where an external bias resistor is used with a transmitter, increasing the resistance (and decreasing the signal current and optical power) will increase reliability.

Because device characteristics degrade and reliability decreases with increasing temperature, most transmitters are designed to operate only over the commercial temperature range. Typical maximum operating temperatures are 70°C, and lower temperature limits are typically 0°C. Transmitters for military temperature limits (–55 to +125°C) are also available.

4.11: LED vs. Laser: System Tradeoffs

The choice of an LED or laser will have significant impact on link performance. The design of the bias and input circuits will also impact performance. For example, it is a waste of money to have a 10 Gbit/sec laser driven by a TTL (15 Mbit/sec) digital gate. In general, the lowest cost component which meets transmitter specifications should be used. (Cost must be evaluated over the life of the system, including any replacement costs due to lower reliability.)

For analog applications, the LED is the preferred optical source. As discussed previously, a laser diode typically exhibits kinks and other

---

[1] For example, the package should not be cooled below its minimum operating temperature specification.

non–linearities. The LED has a more linear P–I characteristic, and is less sensitive to temperature variations.

The modulation bandwidth of an LED can be controlled by controlling the bias point. For an LED, for example, the bandwidth can be increased by increasing the DC bias current. In this case:

$$P \Delta f = K I$$

where P is the optical output power, Δf is the 3 dB modulation bandwidth, and K is a constant that depends on the specific LED device. Note that there is no direct relationship between the carrier frequency (LED wavelength) and the modulation bandwidth. There is a limit to the bandwidth that can be achieved, since the device reliability decreases with the increased bias current.

The transient (step) response of an LED is influenced by the diode junction capacitance C_j. Typical values for C_j range from 300 to 1000 pF. The turn–on delay time, $t_{1/2}$, and risetime, t_{10-90} are given by:[1]

$$t_{1/2} = [2\,k\,T/q]\;[C_j/I_p]\ln(I_p/I_s) + \tau\ln(2)$$

$$t_{10-90} = (C_j/I_p)(4\,k\,T/q)\;\ln(9) + \tau\;\ln(9)$$

where T is the temperature in Kelvin, I_p is the driving current pulse, I_s is the diode saturation current and τ is the carrier (recombination) lifetime. Typical values for τ are around 1–10 nsec. I_s is determined from the diode equation,

$$I = I_s\,e^{(qV/nkT)}$$

For an LED, typical values of I_s are anywhere from 10^{-20} to 10^{-15} A, and typical values of n between 1 and 2. Clearly, the switching speed increases as the current pulse amplitude increases. For this reason, a brief current spike may be used on the rising and falling edges to increase the switching speed, similar to current peaking used in switching fast bipolar transistor signals.

The delay and risetime depend on the same device characteristics, but in slightly different fashion. It is possible for the delay time to be longer than the risetime, but it is also possible for the delay time to be shorter than the

[1] Note that some data sheets give the 20–80% transition time instead of the 10–90% time.

risetime. The first term in the delay is the time required to charge the junction capacitance, and the second term is due to the inherent delay between current flow and optical emission due to carrier transition times in the semiconductor material. For small modulation current and low bias currents, the delay time is larger than the risetime. For large switching currents, both times decrease. Since the delay time decreases faster than the risetime, there is a drive current at which the delay becomes equal to the risetime; further increases in drive current then make the delay time smaller than the risetime.

EXAMPLE:

An LED has a junction capacitance of 300 pF, a junction leakage current of 150 nA, and a carrier (electron–hole) lifetime of 3 nsec. It is operated with a peak current of 25 mA at room temperature. Compare the delay and risetimes.

$$t_{1/2} = [300\text{ pF}/25\text{ mA}][(2)(1.381 \times 10^{-23}\text{ J/K})(298\text{ K})/1.6 \times 10^{-19}\text{ C})] \times [\ln(25\text{ mA}/150\text{ nA}) + (3\text{ nsec})\ln(2)]$$

$$t_{1/2} = 9.5\text{ nsec}$$

$$t_{10\text{–}90} = [300\text{ pF}/25\text{ mA}][(1.381 \times 10^{-23}\text{J/K})(298\text{ K})/1.6 \times 10^{-19}\text{C})]\ln(9) + (3\text{ nsec})\ln(9)$$

$$t_{10\text{–}90} = 2.9\text{ nsec}$$

Figure 4-17 illustrates characteristic transitions. The four curves shown illustrate the possible combinations of long and short delays and fast and slow risetimes.

Laser diodes have higher modulation bandwidths than LEDs, with commercial laser diodes covering ranges between 10 Mbit/sec and 15 Gbit/sec. The modulation bandwidth of a particular laser diode is dependent on the device structure and the operating current. The modulation bandwidth is usually limited mainly by relaxation oscillations (self–resonance). Lasers may also show oscillations in the few GHz range due to interactions between the optical power in the cavity and the facet mirrors or semiconductor material: at high optical–power densities, defects are created, causing the gain to decrease. This in turn causes the power density to decrease, and the defects "heal" themselves. Very high optical–power densities may permanently damage the facets or the semiconductor materials, leading to a permanent loss of cavity gain, usually observed as an increase in threshold current.

The risetime of a transmitter module may be significantly slower than that of the optical diode unless care is taken in the electrical design and physical layout of the module. The main circuit considerations are RC time constants (including the bias resistor–junction capacitance time constant), and the ability of the driver to source and sink the LED or LD current as necessary. Layout of components adds wiring capacitance and wiring inductance to the circuits (e. g. a wire bond induces approximately 1 nH of inductance).

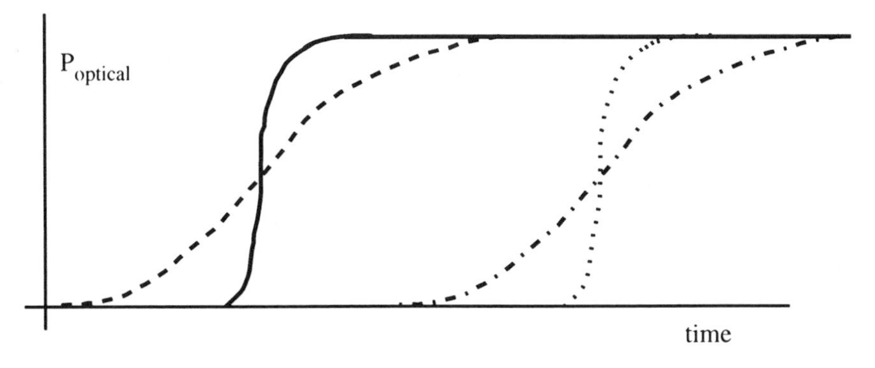

Figure 4-17: Transition Characteristics.

With the present trend to purchasing transmitters as complete modules, the designer is relieved of many of the layout constraints. For high–speed systems, however, care must be taken to keep the wires between transmitter electronics and laser diodes as short as possible, as well as wires between modules.

Lasers will provide higher link bandwidth than LED–based links if the optical source speed is a significant bandwidth limitation. Optical diode speed must be separated from electrical speeds, since both LEDs and lasers are subject to RC time constant effects.

Once an optical signal has been coupled into a fiber, the signal undergoes pulse broadening due to interactions between the optical wave and the fiber waveguide. One of the pulse broadening mechanisms is directly proportional to the signal (and hence source) spectral width. When this is a significant bandwidth limitation, a laser will provide improved link bandwidth due to the smaller spectral width of the laser diode.

4.12: Summary

A transmitter consists of electronic bias and signal conditioning circuits and an optical source diode. Transmitter circuits must provide suitable DC bias. The optical source diode may be a light–emitting diode or a laser diode.

Multiplexing circuits may be included on the transmitter or as a separate board or module. All of the standard electronic multiplexing schemes may be used, including time domain and frequency domain multiplexing. In addition, wavelength division multiplexing can be used for optical carriers.

LED and laser diodes depend on the generation of photons due to electron–hole recombination. The wavelength of the emitted light depends on the semiconductor material, while the spectral width and modulation capabilities are determined by the device structure and the bias network.

Both LED and laser diodes are biased in the forward direction. A DC bias current should be provided for the laser diode. Laser diodes also require thermal management, since the required threshold current varies strongly with temperature.

A laser provides a higher link performance than an LED, due to its higher modulation speed, smaller spatial spreading, and smaller spectral width. An LED provides better linearity and higher reliability. An LED also costs less than a laser diode, with the exception of the laser used for compact disc players.

4.13: Exercises

1. Assuming that a 1 μm optical source could be modulated at 1% of the carrier frequency, how many 5–MHz TV channels could be sent using FM modulation of this carrier?
2. Draw a block diagram of an electronic system that multiplexes 16 lines down to 1 line, using four–way multiplexers.
   (a) How many levels (cascaded stages) of MUXing are required?
   (b) If the data rate is 10Mbit/sec on each input line, what are the required data rates at each level of multiplexing?
3. What is the coding efficiency of the T–1 carrier?
4. Describe the differences between gain guiding and index guiding.
5. An LED is made from GaAs, with a bandgap energy of 1.424 eV. What are the wavelength and frequency of the optical carrier?
6. An LED has a wavelength of 1.55 μm. What is the bandgap of the semiconductor material?

7. Calculate the leakage current I_s of an LED which carries 20 mA at 1.1 V if $n = 1.2$.
8. Plot the delay time and risetime vs. peak current (using the same graph) from 5–50 mA. The LED has a junction capacitance of 750 pF, a leakage current of 2×10^{-17}A, and a carrier lifetime of 2 nsec.
9. An LED diode has a junction capacitance of 2700 pF, a leakage current of 2 pA, and a carrier lifetime of 5 nsec. (a) Plot the delay time and risetime vs. peak current from 100-500 mA. (b) At what current are the delay and risetimes equal?
10. A laser diode has a 50 mA threshold current at 55°C and 25 mA at 30°C. Determine I_z and T_z.
11. A laser diode has a threshold current which increases 25% as the temperature is increased from 30°C to 50°C. Determine T_z.
12. A laser diode has a 35 mA threshold current at 55°C and 15 mA at 30°C. Determine the threshold current at 25°C.
13. A laser diode has an $I_z = 0.05$ mA and $T_z = 132$ Kelvin. Plot the threshold current vs. temperature for 0-70°C.
14. An LED has a MTTF of 25 years at 25°C and 10 years at 55°C. What is the activation energy of the failure mechanism?
15. An LED has a MTTF of 20 years at 25°C. The failure mechanism has an activation energy of 0.69 eV. What is the MTTF at 55°C?
16. An LED has a MTTF of of 20 years at 25°C and 10 mA. What is the MTTF when the current is increased to 50 mA for $n = 1.5$? The MTTF at 50 mA for $n = 2$?
17. What are the advantages and disadvantages of using bias current control vs. temperature regulation to control the operating threshold for a laser diode?
18. An LED has a modulation bandwidth of 10 MHz at a bias current of 30 mA. What is the modulation bandwidth if the current is decreased to 15 mA?
19. Discuss the tradeoffs between modulation bandwidth and reliability in a diode. What are the relative advantages of low and high current bias levels?

Chapter 5 Receiver Modules

Receivers convert optical inputs back into electrical inputs. A photodiode converts the electrical signal into a current signal. Because the received optical signal and resulting electrical current have small amplitudes, the receiver will usually contain one or more amplification stages. The receiver may also contain filters or equalizers to improve signal quality. A digital receiver converts the amplified signal into a digital signal level.

A receiver is generally designed to be used with a particular transmitter. For example, a TTL transmitter is generally paired with a TTL receiver. It is difficult to mix receivers with different transmitters, particularly from different manufacturers, because the information is not always encoded in the same fashion. Even if the receiver is designed to meet the same electrical interface specifications (e. g. TTL levels), the information may be encoded in different ways (e. g. one manufacturer may use NRZ and another Manchester within the transmitter). Only when both electrical *and* optical encoding and timing follow a common standard can inter–operability be achieved.

5.1: Receiver Overview

A receiver consists of a photodiode, bias resistor, and low–noise pre-amp, as shown in Figure 5-1. It may also contain additional amplification and filters or equalizers. These components may be on a single integrated circuit, a hybrid, or a printed circuit board.

A complete receiver system may incorporate a number of support functions. One of the most common is clock recovery. Clock recovery circuits extract a clock (timing) signal from the bit stream, so that output information can be clocked consistently. This helps to preserve framing information and decreases the BER. Clock recovery is necessary in fiber optic systems, since the clock is not transmitted separately from the data.

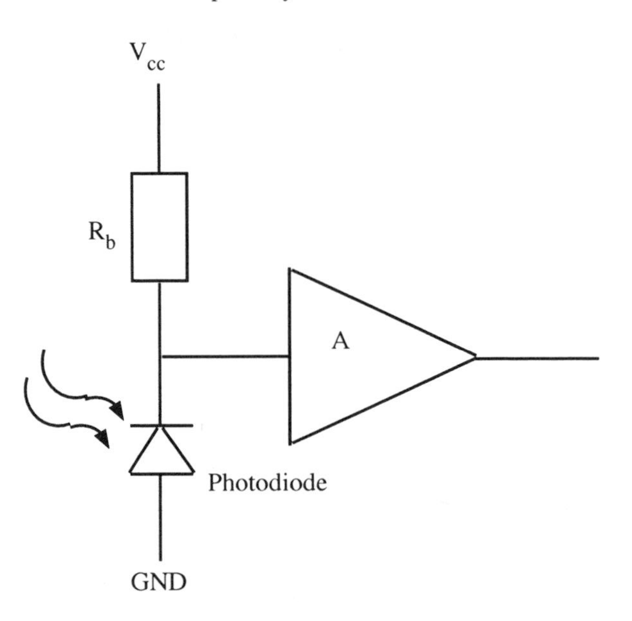

Figure 5-1: Receiver Block Diagram.

Other support functions may include decoding (e. g. of 4B/5B encoded information), error detection and error recovery, and detection of link failures (i. e. loss of optical signal or loss of modulation). A complete receiver system usually incorporates a number of ASICs and occupies a printed circuit board.

A receiver must have high sensitivity, high bandwidth (or fast risetime), and low noise. The photodetector must have small size and be compatible with the wavelength of the system transmitter. The receiver must provide high gain and a large output signal with low bias complexity, high reliability, and low cost. This is achieved by using a diode detector and integrated circuit technology for the amplifier and support circuits.

5.2: Device Structures

The photodiode converts the optical signal directly into electrical current, using the reverse of the physical process in the LED. When a photon of sufficient energy enters a semiconductor material, it may be absorbed by that material. Photon absorption results in the freeing of an electron from a chemical bond, resulting in a hole–electron pair. Both the hole and electron are free carriers, and can move within the semiconductor in the presence of an electrical field.

The minimum energy that the photon must have to create a hole–electron pair is the semiconductor bandgap energy E_g. If the photon has more energy than this, the hole and electron will convert the excess energy into kinetic energy (heat). A photon must have at least $2E_g$ of energy before it can create a second pair. Unlike emitting semiconductors, the absorbing semiconductor does not need to be a direct bandgap material. Indirect bandgap materials are most often used. In particular, silicon is used for the short–wavelength window.

A photodiode is characterized by the window in which it can be used. The cutoff energy has a corresponding cutoff wavelength. Silicon photodiodes, with a cutoff wavelength of 1100 nm, can only be used in the short–wavelength window. Photodiodes for use in the long wavelength window (1200–1600 nm) are more commonly made from III–V materials, although germanium (Ge) is suitable for some applications.

Silicon photodiodes have a number of advantages. Silicon wafers are commercially available up to 8 in. across, compared to 4 in. for III–V materials such as GaAs; and silicon IC processing is a mature industry. Silicon photodiodes can be easily integrated with transistors on an IC, making it possible to fully integrate the receiver. A fully integrated circuit produces the lowest overall cost and improves reliability.

EXAMPLE:

Silicon has a bandgap energy of 1.12 eV. What is the minimum energy of photons that can be absorbed by a photodiode? What is the maximum wavelength of the photons?

$E_{min} = E_g = 1.12 \text{ eV}$

λ_{max} occurs at E_{min}

$\lambda_{max} = hc / E_g = (1.240 \text{ eV–}\mu\text{m}) / (1.12 \text{ eV}) = 1.11\ \mu\text{m}$

There are two types of photodiode structures: the PIN and APD. The PIN structure is similar to the basic p–n junction diode, except that an intrinsic layer is sandwiched between the p and n regions. The APD structure is more complex, having several p and n layers.

Figure 5-2 shows the basic structure of the two types of photodiodes. The PIN photodiode is named for its structure of p–i–n layers. In both the PIN and APD photodiode structures, the intrinsic layer is relatively thick (on the order of 5–10 μm), enhancing photon capture.

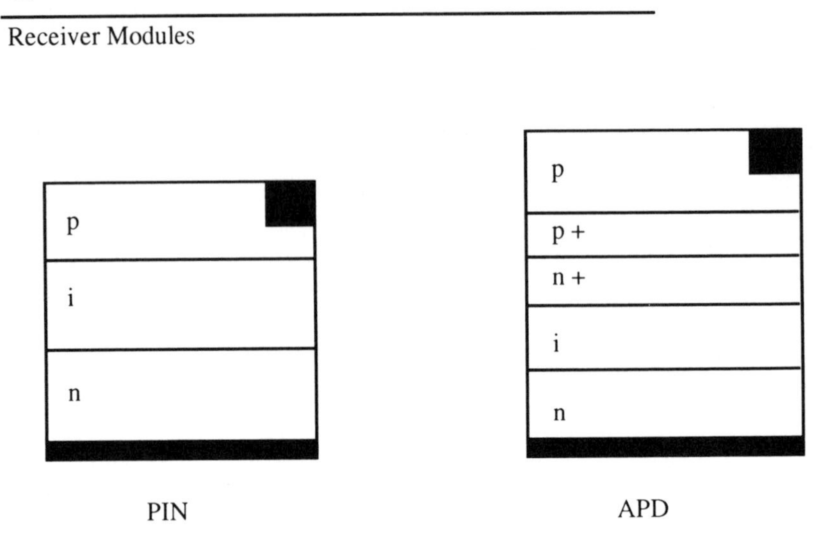

Figure 5-2: Photodiode Structures.

The probability of a photon being captured within a region of semiconductor material depends on a number of factors, including the photon energy and the semiconductor material properties. It also depends on the thickness of the absorbing layer, with a greater thickness absorbing more light.

Figure 5-3 shows the I–V characteristics of a PIN photodiode. As the optical power is increased, the reverse current increases. Photodiodes are operated in the reverse bias region, since the signal current is on the order of 10 nA to 100 μA. A forward current of 1 mA or more would totally swamp this signal current, making it impossible to separate the signal from the noise.

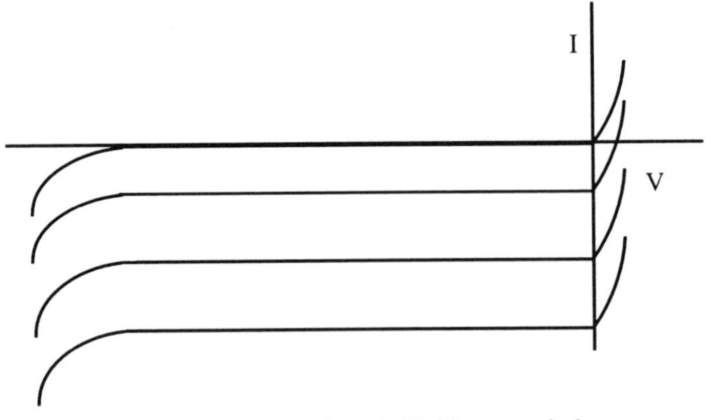

Figure 5-3: Photodiode I–V Characteristics.

When no light shines on the photodiode, only the leakage current of the junction flows. This is referred to as the “dark current”. Dark currents are typically on the order of a 1–10 nA. Photocurrent, caused by the optical signal, adds to the leakage current. In an APD photodiode, the dark current increases with the bias. Most devices can be characterized as having dark current:

$$I_d = M\,I_{db} + I_{ds}$$

where I_{db} and I_{ds} are measured at low bias (where M = 1).

The dark current depends on the reverse bias, on the temperature, and on the semiconductor material. For analog applications, it is important to select a semiconductor material having low dark currents and low bias and temperature sensitivity. Silicon is the most commonly used material, since it has a low dark current and acceptable bias sensitivity for many applications. Germanium has the lowest bias sensitivity, but has a dark current that is almost 10^4 times larger. Photodiodes made from III–V compound semiconductors will generally exhibit dark currents somewhere between those of silicon and germanium. They will also generally exhibit more variation in dark current with bias.

An optical signal will cause a current in the photodiode as long as the photon energy is above the photodiode's semiconductor bandgap energy. This photocurrent is added (as an n–to–p current) to the dark current. This is true independent of the diode's bias, as shown in Figure 5-3. Typical photocurrents are on the order of 1 μA, which is larger than the dark current and smaller than the forward current of 1 mA or more. Therefore a photodiode is operated in the reverse bias region to improve the SNR.

Photodiode switching speed increases with increasing reverse bias because the velocity with which carriers (holes and electrons) cross the intrinsic layer is dependent on the electric field in that layer. The carrier speeds will reach a limit at around 10^4 V/cm in silicon. This limiting electric field strength is called the "critical field", and is dependent on the semiconductor material. For example, in a silicon photodiode with a 10 μm intrinsic layer depth, the critical field is reached at a bias of

$$V = E\,L = (10^4 \text{ V/cm})\,(10^{-3} \text{ cm}) = 10\text{V}.$$

Most PIN photodiodes are designed to reach critical fields at 5-15 V reverse bias. Increasing the bias above this voltage produces relatively smaller increases in response speed.

The APD photodiode, with its more complex structure, amplifies the signal current within the photodiode itself. The two heavily doped layers (p+/n+) produce junction breakdown at relatively low voltages. In this avalanche breakdown region, the carriers accelerate sufficiently to generate additional carrier pairs though collisions. Figure 5-4 shows the avalanche region where one photon–generated hole has resulted in three collisions in the breakdown field region. In each collision, one hole comes in and two holes and one electron leave.[1] The distance a carrier travels before a collision occurs and the total number of carriers generated per photon–generated carrier will vary due to the statistical nature of the collision process.

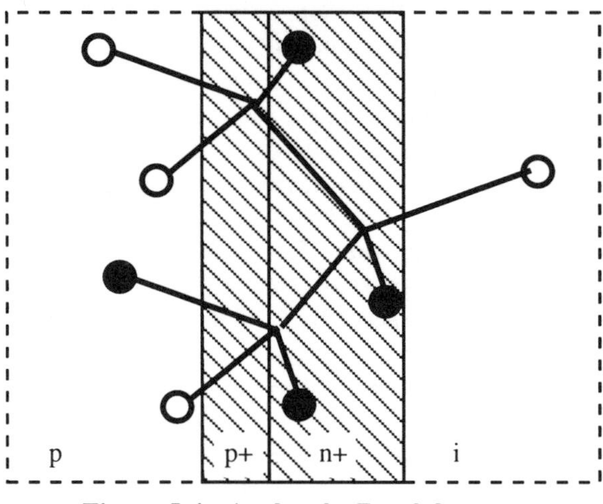

Figure 5-4: Avalanche Breakdown.

In avalanche breakdown, both holes and electrons may collide. The gain M is the ratio of the average number of carriers leaving the avalanche region to the number that enter. For Figure 5-4, M = 3. The gain will vary as a function of bias voltage, temperature, optical wavelength, and semiconductor material. It is possible to operate devices with gains as high as 1000, but practical gains are usually on the order of 5–50.

If the reverse voltage is increased too far, the photodiode enters second breakdown, causing damage to the device. A large bias may also cause the diode gain to "saturate", where a photon will initiate a large current flow which is then self–sustaining, resulting in a one–shot detector.

An APD achieves gain only when the bias exceeds the first breakdown voltage (V_{br}). Below the avalanche breakdown voltage, M = 1. For most practical devices, V_{br} is between 25 and 200 V. Switching speed is determined mainly by the physics of the avalanche region. Risetime is very fast when the gain is high (since a single photon can result in a large output current).

[1] Or one electron comes in and two electrons and one hole leave.

However, fall times increase with increasing gain, since it takes longer for the avalanche region to become free of collision generated carriers. Near second breakdown, it becomes impossible to turn the APD off, so that one photon–generated carrier can saturate the gain mechanism.

Because the APD must be biased above V_{br} to achieve gain, an APD cannot be used on a 5 V printed circuit board. Use of an APD requires a stable power supply, which can add to system cost if such a power supply is not already available. The more complex bias circuits required for APDs increase system cost and decrease reliability. The APD itself is also less reliable than PIN photodiodes, particularly if biased near second breakdown.

In most local area network links, the PIN is the preferred device. This is mainly because it can be operated from a standard board power supply (typically between 5 and 15 V). APD devices, while providing 5–10 dB more sensitivity and half the risetime, are generally reserved for long–haul telecommunications links, where the increased repeater spacing provides cost savings to overcome the costs associated with special power supplies and decreased reliability. For either type of photodiode structure, the amount of photocurrent I_p generated by a photodiode is given by:

$$I_p = \mathrm{M}\, \boldsymbol{R}\, \mathrm{P}$$

R is the responsivity, and it is a function of both the semiconductor material and the optical wavelength. Typical values of ***R*** range between 0.2 and 0.8 A/W. Because the incident optical power and resulting current are small, ***R*** is frequently expressed in μA/μW.

EXAMPLE:

A silicon PIN diode has a responsivity of 0.58 A/W at 830 nm. Optical power of 800 nW at 830 nm falls on the diode. What is the resulting photocurrent?

Since a PIN diode has no internal gain, M = 1.

$I_p = \mathrm{M}\, \boldsymbol{R}\, \mathrm{P}$ = (1) (0.58 A/W) (800 nW) = 464 nA

EXAMPLE:

A silicon APD diode has a low–bias responsivity of 0.58 A/W at 830 nm, and is operated with a gain of 5. Optical power of 800 nW at 830 nm falls on the diode. What is the resulting photocurrent?

$I_p = \mathrm{M}\, \boldsymbol{R}\, \mathrm{P}$ = (5) (0.58 A/W) (800 nW) = 2320 nA = 2.320 μA

The responsivity ***R*** exhibits a strong dependence on the optical wavelength. In particular, the responsivity drops to zero above the cutoff wavelength. Figure 5-5 shows how responsivity varies with wavelength for silicon.

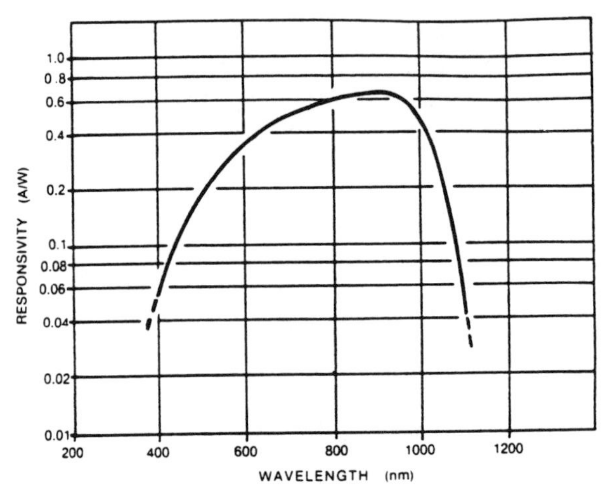

Figure 5-5: Responsivity vs. Wavelength for Silicon.

Courtesy of AMP Inc.

As can be seen from the examples, the photocurrent is on the order of several nA to a few μA for many links. A short link may have a current as high as a few mA. These currents are small, and will usually not drive following circuits (e. g. digital gates, telephone headsets). In most links, an amplifier is an integral part of the receiver.

5.3: Amplifiers

The first amplifier (pre–amp) in a receiver must have very low input noise, to avoid swamping the small signal current provided by the photodiode. The photodiode, its bias resistor, and the pre–amp should be designed as a unit.[1] Interactions between these three components will determine receiver bandwidth, SNR, and sensitivity (minimum detectable signal).

Amplifiers can be divided into four groups, based on the input transistor

---

[1] This section focuses on key factors in amplifier selection. Those designing their own amplifiers are referred to the bibliography (e. g. the book by John Gowar).

type (FET or BJT) and configuration (transimpedance or high impedance).[1] All four types are used in receiver design, depending on the desired characteristics. Performance of the four types are illustrated in Figure 5-6. The quantum limit is the sensitivity (minimum detectable power) due only to the statistical nature of photon detection and carrier pair generation.

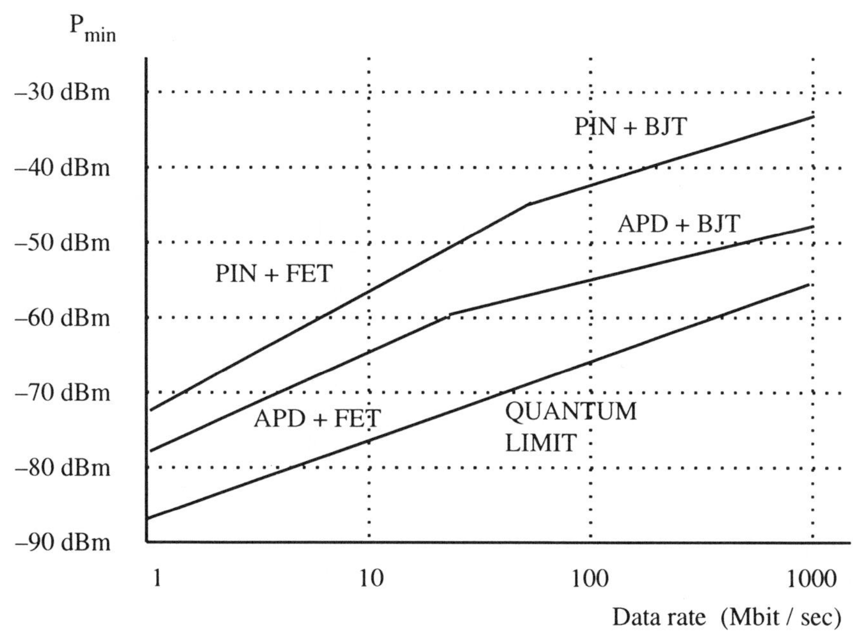

Figure 5-6: Pre–amp Characteristics.

EXAMPLE:

A receiver containing a PIN photodiode and an FET amplifier has the characteristics shown in Figure 5-6. Determine the sensitivity at 10 Mbit/sec.

From the Figure, $P_{min} = -56$ dBm

$$P_{min} = 1 \text{ mW } [10^{(-56/10)}] = 2.51 \text{ nW}$$

As seen in the above figure, the APD provides greater sensitivity at all data rates. A FET amplifier provides improved sensitivity at lower data rates,

[1] These may also be referred to as feedback and open–loop configurations.

a bipolar amp at higher. The choice of amplifier type will therefore be influenced by the desired operating frequency range.

A high–impedance pre–amp, such as that shown in Figure 5-7 (a), exhibits a high gain. A high gain implies that the amplifier will saturate for a low input signal. To achieve a higher dynamic range, a lower gain is required. This is achieved by using an amplifier with negative feedback, as shown in Figure 5-7(b). In addition to increased dynamic range, the feedback pre–amp has decreased sensitivity and wider bandwidth.

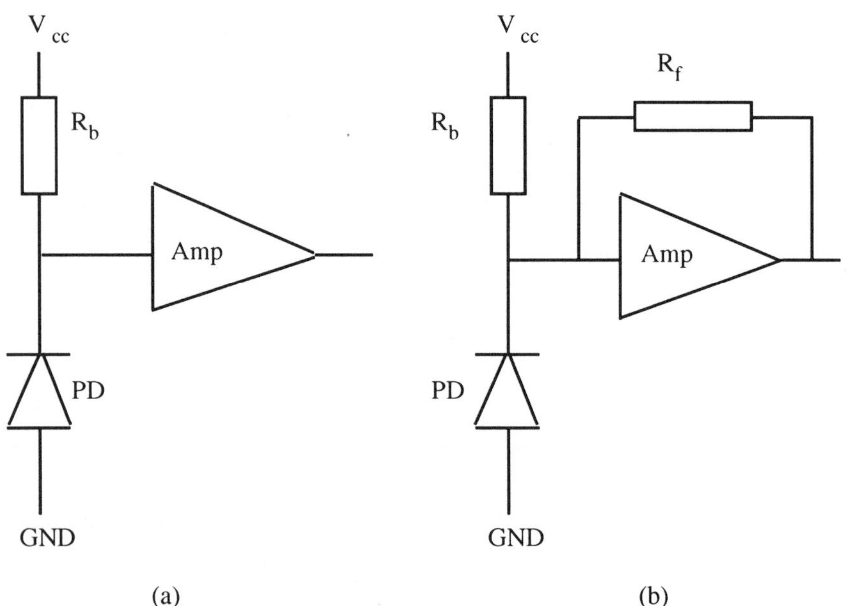

(a)
High–impedance

(b)
Transimpedance

Figure 5-7: Amplifier Stages.

The bandwidth for a transimpedance amplifier is:

$$A / [\,4 R C\,]$$

where A is the DC amplifier gain, and R and C are the input node resistance and capacitance, respectively. For a high–impedance amplifier, the bandwidth is:

$$1 / [\,4 R C\,]$$

From these equations, it can be seen that the bandwidth of the transimpedance amplifier is A times greater than for the open–loop design. The dynamic range is increased by a factor of AR. Table 5-1 lists some of the major tradeoffs for the two types of amplifiers.

Once the photodetector and pre–amp have been selected, the two must be connected. This can be done using DC coupling, as shown in Figure 5-7. DC coupling is required if the information is not run–length limited, either by the transmitter or other system modulation. This is the case for some sensor systems and for NRZ transmitters with no system information bandwidth constraints.

Table 5-1: Amplifier Configuration Tradeoffs

High Impedance

Advantages:	*Disadvantages:*
Higher sensitivity @ low f	Lower bandwidth
Higher gain	Lower dynamic range
	Integrates the signal
	FET input noisy at low f (<10 kHz)

Transimpedance

Advantages:	*Disadvantages:*
Higher bandwidth	Lower sensitivity (about 2–3 dB less)
Wider dynamic range	More components
Higher sensitivity @ high f	
Improved linearity	
Low output impedance	
Controllable gain (R_f)	
Stable gain	

If the information is bandwidth limited, it is desirable to AC couple the photodiode and the pre–amp (Figure 5-8). This eliminates low frequency noise and drift effects from being coupled into the signal. For example, FET transistors are also known to have high 1/f noise,[1] sometimes large even up

[1] So called because the noise power varies as (1 / f).

into the audio region (to 10 kHz for some devices). AC coupling blocks low frequencies, and eliminates low frequency noise.

In a digital receiver, the output of the amplifier must be quantized (converted to a logical "1" or "0" level). In a high–impedance amplifier, there is capacitive coupling from the output back to the input due to transistor capacitance. This capacitive feedback causes the pre–amp to integrate the signal. AC coupling provides cancellation of this effect, since the DC level is forced to "drift" back to zero.

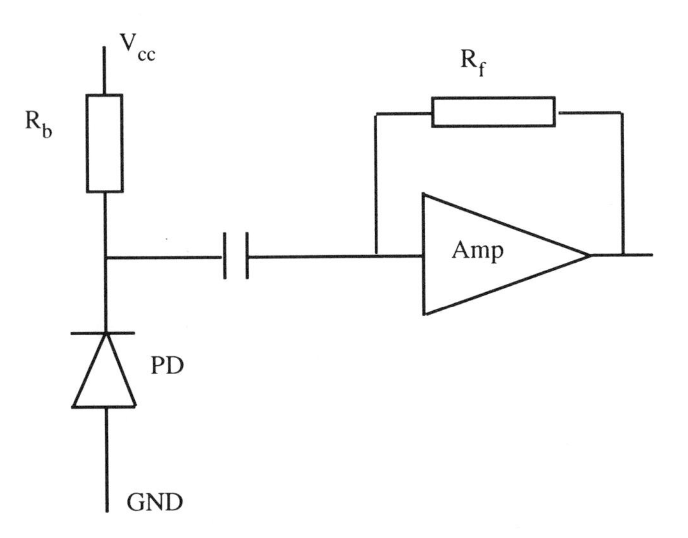

Figure 5-8: AC–coupled Transimpedance Pre–amp.

AC–coupled receivers are more commonly used than DC–coupled receivers. Reducing the noise improves sensitivity, improving the receiver performance. The main consideration in selecting an AC–coupling capacitor is that the receiver must not have a low frequency cutoff in the same frequency range as the signal. The information must be bandwidth limited if such a receiver is to be used, and receivers are not interchangeable if they have different cutoff frequencies.

5.4: Receiver Bandwidth

The receiver bandwidth is determined by the photodiode, the coupling capacitor (if any), the pre–amp, and any following filter or equalizer. Circuit analysis follows the same techniques as for other electronic circuits. Figure 5-9 shows the AC–equivalent circuit for an AC–coupled receiver.

The input to the AC–equivalent circuit is the AC photocurrent i_p. There are two parasitic capacitors, C_j and C_a, for the photodiode and pre–amp input, respectively. The parasitic resistor R_a is the input resistance of the pre–amp. The receiver has a cutoff frequency of the lower of the filter and pre–amp RC input bandwidth cutoffs.

Typical values for C_a are around 1–100 pF, while values for C_j are around 10–1000 pF. R_a is typically around 10 MΩ or more for FET input pre-amps and above 10 kΩ for bipolar input pre–amps, depending on the design. In analyzing the high frequency cutoff, the coupling capacitor C_c can generally be ignored, while the two parasitic capacitors can be ignored in determining the low frequency cutoff.

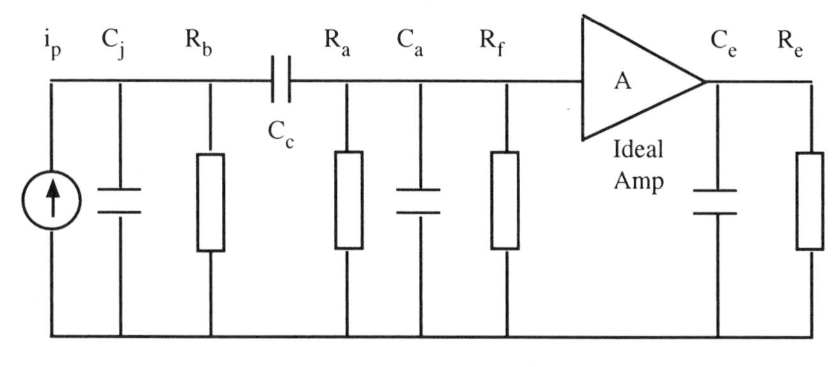

Figure 5-9: AC–Equivalent Circuit.

EXAMPLE:

A receiver is designed to handle data rates from 10 to 40 Mb/sec. What is the frequency range of the permitted signals? Assume only the first harmonic is to be retained.

$$f_{min} = \text{DR} = 5 \text{ MHz}$$
$$f_{max} = (2)\,[40 \text{ Mbit/sec}] = 80 \text{ MHz}$$

The next two examples illustrate how RC time constants in the receiver electronics can limit the receiver bandwidth. Care must be taken to minimize the wiring capacitance as well as to select amplifiers with high bandwidth, high input impedance, and low input capacitance. In most designs below 10 MHz, the electronics will dominate the receiver bandwidth. Photodiode risetime becomes significant at higher speeds.

EXAMPLE:

A receiver has a bias resistor of 10 kΩ and a photodiode capacitance of 4 pF. It is DC coupled to a high–impedance pre–amp with an input impedance of 10 MΩ and input capacitance of 8 pF. The output filter has a frequency cutoff of 10 MHz. Determine the receiver bandwidth.

a) The input stage capacitance is:

$$C = C_j + C_a = 4\text{ pF} + 8\text{ pF} = 12\text{ pF}$$

The input stage resistance is:

$$R = R_b \,||\, R_a = 10\text{ k}\Omega \,||\, 10\text{ M}\Omega = 9.99\text{ k}\Omega$$

$f_{RC} = 1 / [\, 2\,\pi\, R\, C\,] = 1.33$ MHz.

b) The receiver cutoff frequency is the lower of the two:

$f_{max} = \text{MIN}\,(1.33\text{ MHz}, 10\text{ MHz}) = 1.33\text{ MHz}$

EXAMPLE:

A receiver has a bias resistor of 10 kΩ and a photodiode capacitance of 4 pF. It is DC coupled to a transimpedance pre–amp with an input impedance of 10 MΩ, input capacitance of 8 pF, and a 10 kΩ feedback resistor. Determine the receiver bandwidth.

a) The input stage capacitance is $C = C_j + C_a = 4\text{ pF} + 8\text{ pF} = 12\text{ pF}$
The input stage resistance is $R = 10\text{ k}\Omega \,||\, 10\text{ M}\Omega \,||\, 10\text{ k}\Omega = 5.00\text{ k}\Omega$
$f_{RC} = 1 / [\, 2\,\pi\, R\, C\,] = 2.65$ MHz

b) The receiver cutoff frequency is the lower of the two:

$f_{max} = \text{MIN}\,(1.33\text{ MHz}, 10\text{ MHz}) = 2.65\text{ MHz}$

Filtering (or equalization) is used after the amplifier for a number of reasons. First, the fiber optic link (particularly the transmitter and receiver) introduce non–linearities into the information transfer function. Non–linearities cause information transmitted at one AC frequency to also appear at harmonics of that frequency, and to cause signals at different frequencies ω_1 and ω_2 to become coupled at frequencies $\omega_1 + \omega_2$ and $|\omega_1 - \omega_2|$. These effects are referred to as harmonic and inter–modulation distortion, respectively. Selective filtering of the output can remove harmonic and inter–modulation signals that are not in the same band as the original information. This filtering can also remove low and high frequency noise.

5.5: Receiver Noise Performance

The receiver is the major contributor to noise in a fiber optic link. The transmitter generally provides a high SNR,[1] and the transmission medium does not couple in external (EMI) noise. At the receiver, the noise amplitude is a significant fraction of the signal amplitude, and more attention must be given to circuit design and component selection.

Figure 2-7, shown again in Figure 5-10, shows the eye pattern of the transmitted and received signals for a 10 Gbit/sec link. The increase in "fuzziness" of the received signal is due to the noise in the link, primarily caused by the receiver. An eye pattern, created by sending a pseudo–random set of data bits, is often used in characterizing receiver and link noise performance.

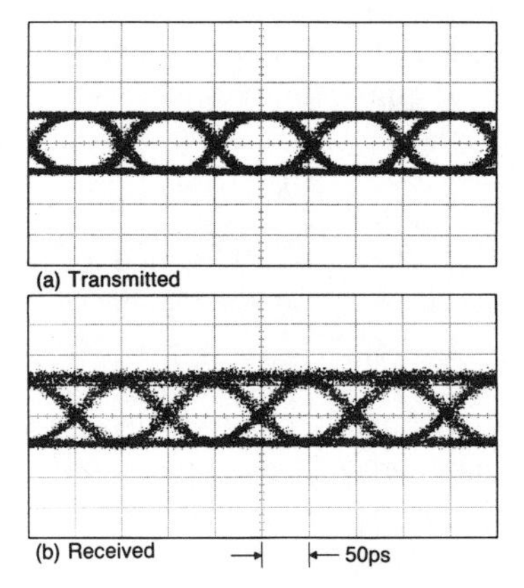

Figure 5-10: Eye Patterns at 10 Gbit/sec.

Reprinted from A. Gnauck, C. Burrus, and D. Ekholm, "A Transimpedance APD Optical Receiver Operating at 10 Gb/sec", *IEEE Photonics Technolgy Letters*, May 1992. Copyright 1992 by IEEE. Used with permission.

There are three major sources of noise at the receiver: fluctuations in the optical input, the photodiode, and the amplifier input electronics (including the bias resistor). Noise from the optical input is produced because the optical

[1] With the exception of laser–induced reflection and modal noise.

input is not continuous, being composed of discrete photons. This noise results in a fundamental quantum limit on the noise, as seen in Figure 5-6. Photon quantum noise is generally not a problem for fiber optic data links, since it is several dB down from the other major noise sources.

A quantum process such as photon arrival, current flow across a junction, or avalanche gain, will result in noise. Consider the case of current flow. If the average current flow past a junction is 1 nA, then an average of 6.242×10^9 electrons will cross the junction in each second. This is the same as an average of 6.242 electrons per nanosecond. In some 1–nsec intervals, an observer will see 6 electrons. In other intervals, there may be 4 or 5 electrons, or 7 or 8 or more, or even no electrons at all. The distribution of all observations follows a classic Poisson distribution:

$$P_n = N^n e^{(-N)} / n!$$

where P_n is the probability of observing n items (electrons) in an interval where the average is N. The probability of observing between a and b number of items in an interval is given by:

$$\sum_{a}^{b} P_n$$

EXAMPLE:

An avalanche photodiode is biased to have a gain M = 3.5. For one incident carrier into the avalanche region, (a) what is the probability of exactly 3.5 collisions occurring? (b) what is the probability of 2 or fewer collisions occurring?

(a) P = 0, since the number of actual collisions must be an integer.

(b) $P_n = 3.5^0 e^{(-3.5)} / 0! + 3.5^1 e^{(-3.5)} / 1! + 3.5^2 e^{(-3.5)} / 2!$
$P_n = (1)(0.03)/1 + (3.5)(0.03)/1 + (12.25)(0.03)/2 = 0.32$ or 32%

Quantum statistics are important in fiber optic receiver noise analysis. Consider a 100 Mbit/sec data link in which the received photocurrent is 10 nA. In the bit period of 10 nsec, the average electron flow is 62.4 electrons. The fluctuation is on the order of $N^{-0.5}$, which is 8 electrons or almost 13%. Clearly, any signal swing must be significantly larger than this to achieve a high SNR or good BER.

The photodiode contributes noise through its current (both dark and photo currents), and through fluctuations in the avalanche process. Current, consisting of a flow of discrete electrons, will fluctuate in a quantum manner; this is often referred to as "shot" noise. Similarly, the avalanche mechanism consists of a discrete number of collisions for each photogenerated carrier.

The mean square current noise associated with the photodiode is:[1]

$$< i_{\mathrm{np}}^{2} > \; = \; 2\,q\,B\,(\,I_p + I_{db}\,)\,\mathrm{M}^{2+x} \; + \; 2\,q\,B\,(\,I_{ds}\,)$$

where I_p is the photocurrent, I_{db} and I_{ds} are the photodiode leakage current terms, M is the avalanche gain, x is a characteristic of the photodiode,[2] and B is the receiver noise bandwidth.

Bias and feedback resistors contribute Johnson (thermal) noise due to the random motion of electrons. For example, while electrons in a resistor move with an average velocity of a few cm/sec in the direction of current flow, individual electrons move with speeds around 10^5 cm/sec in random directions. The thermal noise associated with a resistor is:

$$< i_{\mathrm{nr}}^{2} > \; = \; 4\,\mathrm{k}\,T B\,/\,R$$

where k is Boltzmann's constant, T is the temperature in Kelvin, B is the receiver noise bandwidth, and R is the resistor value. Resistor noise appears as a current source in parallel with the resistor.

The dark current contribution to noise will depend on bias and on temperature. Leakage currents increase rapidly with increasing temperature. Leakage currents also increase with increasing bias, particularly for the III–V compound semiconductors.

The amplifier also contributes noise. Amplifier noise arises from the shot noise of the input transistors and the thermal noise of any feedback resistor. The feedback resistor has the same thermal noise characteristic described above.

An amplifier is commonly characterized by voltage and current noise power spectral density, e_{n} and i_{n}. RMS values are generally given on data sheets for low–noise pre–amps. The noise contribution of the amplifier is given by:

$$< i_{\mathrm{na}}^{2} > \; = \; 2\,B\,[\,i_{\mathrm{n}}^{2} \; + \; (e_{\mathrm{n}}/\,R)^{2}\,]$$

---

[1] Noise is always described in terms of average power $<i^2>$ or in terms of RMS value. The average value of noise is zero.

[2] x is typically between 0 and 1.0, and is dimensionless.

EXAMPLE:

A PIN photodiode has a dark current of 2 nA and a photocurrent of 50 nA. It is connected to a 10 kΩ bias resistor. The input stage bandwidth is 50 MHz. Compare the photodiode and thermal noise contributions at room temperature (300 Kelvin).

For a PIN, M = 1 and $I_{db} = I_{dark}$.

(a) $< i_{np}^{2} > = 2\,(1.6 \times 10^{-19}\text{Coul})\,(50 \times 10^{6}\text{ Hz})\,(52 \times 10^{-9}\text{A})\,(1) + 0$
$= 0.832 \times 10^{-18}\text{ A}^2$

(b) $< i_{nr}^{2} > = (4)\,(1.38 \times 10^{-23}\text{ J/K})\,(300\text{ K})\,(50 \times 10^{6}\text{ Hz}) / 10\text{ k}\Omega$
$= 82.8 \times 10^{-18}\text{ A}^2$

In this case, the resistor contributes the majority of the noise current.

EXAMPLE:

The photodiode and bias resistor of the previous example are connected to a high–impedance amplifier having noise spectral densities of $e_n = 10$ nV/Hz$^{1/2}$ and $i_n = 6$ pA/Hz$^{1/2}$. Compare the amplifier noise to the thermal noise.

(a) From before, $< i_{nr}^{2} > = 80.8 \times 10^{-18}\text{ A}^2$

(b) $< i_{na}^{2} > = 2B\,[\,i_n^{2} + (e_n / R)^2\,]$
$= 2\,(50 \times 10^{6}\text{ Hz})$
$\times\ [36 \times 10^{-24}\text{ A}^2/\text{Hz} + (10 \times 10^{-9}\text{ V/Hz}^{1/2} / 10\text{ k}\Omega)^2]$

$< i_{na}^{2} > = 3.7 \times 10^{-15}\text{ A}^2$

In this case, the amplifier dominates the total noise current.

The total noise current is given by the sum of the 3 noise terms, assuming uncorrelated noise. The SNR is the ratio of the received power to the noise power. The SNR is given by:

$$\text{SNR} = \frac{(m\,I_p)^2}{< i_{np}^{2} > + < i_{nr}^{2} > + < i_{na}^{2} >}$$

where m is the modulation index ($\Delta I / I_p = \Delta P_o / P_{DC}$), and I_p is the average amplitude of the (modulated) current. The maximum value of m is 0.5.

Another term used to characterize receivers is the noise equivalent power, or NEP. NEP is the noise per unit bandwidth, and has units of $W/Hz^{1/2}$. It is calculated at the amplifier input node with $I_p = 0$:

$$\text{NEP} = \langle i_n^2 \rangle R / B = (\langle i_{np0}^2 \rangle + \langle i_{nr}^2 \rangle + \langle i_{na}^2 \rangle) R / B$$

The SNR and NEP noise calculations are usually made with respect to the amplitude at the input to the amplifier. These may be converted to output voltage noise through the formula:

$$\langle v_{n\,out}^2 \rangle = \langle i_n^2 \rangle R^2 A^2$$

5.6: Environmental Effects

PIN photodiodes are relatively insensitive to variations in temperature and bias voltage. The same cannot be said for APDs. The avalanche gain mechanism is sensitive to many things, including temperature, bias voltage, and optical wavelength.

Figure 5-11 shows how M varies with temperature and bias for a silicon APD. For this photodiode, the first breakdown voltage occurs around 60 V, and shows only slight dependence on temperature. In general, gain increases with decreasing wavelength and with increasing bias voltage, approaching infinity at the second breakdown voltage. The second breakdown voltage increases with increasing temperature. The second breakdown voltage can decrease with age. An APD must be biased so that it does not reach second breakdown over the entire operating temperature range and life.

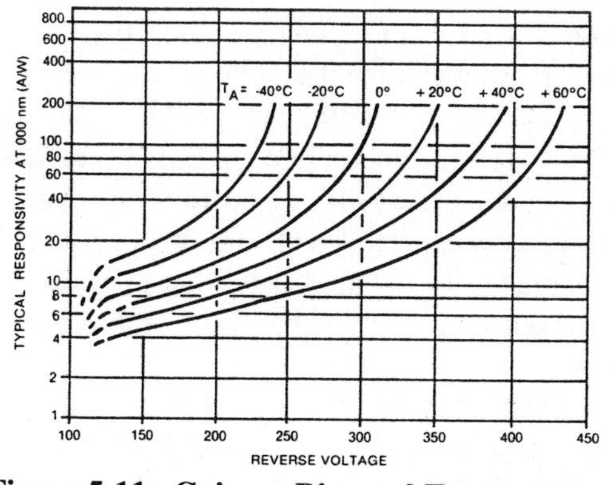

Figure 5-11: Gain vs. Bias and Temperature.

Courtesy of AMP Inc.

Receiver noise increases with increasing temperature due to increased thermal noise and increased leakage currents. Pre–amp gain may also change with temperature. The increase in leakage current can be significant, particularly for semiconductors having smaller bandgaps (those used in the shorter wavelength window).

A second, less critical, temperature effect is seen in the variation of resistor and capacitor values with temperature. Since the absolute values are not critical for most designs, small shifts in value can be tolerated. However, in analog links, the effects of component value tolerances and thermal coefficients must be determined.

Another effect associated with resistors is excess noise. This is additional noise above the thermal noise. Large excess noise is commonly seen in carbon composition resistors. In most receivers, metal film or other low–noise resistors are used.

Reliability decreases with temperature for receiver electronics. In most fiber optic receivers, the receiver circuits will fail before the photodiode, with the exception of the APD biased too near second breakdown.

Finally, there is a unique failure mechanism which is unique to optical systems. In many cases, a system will appear to have lost the optical power into the receiver. This can be caused by a single speck of dust which has gotten into the connection between these components and the fiber.[1] Because dust particles are comparable in size to the fiber diameter and to source and photodiode sizes, one dust particle can block so much light that the SNR decreases or the BER increases dramatically.

5.7: Receiver Layout

Receiver layout is critical in achieving a quality link. Effects which can be tolerated in the transmitter cannot be tolerated in the receiver. There are two particular effects that the board designer must be aware of.

The first effect is coupled noise through EMI or crosstalk. While fiber is a dielectric and does not couple EMI, the board level wiring is done in metal. Each wire acts as an antenna, and can couple small currents into the receiver. For maximum sensitivity, the receiver should be in a metal package to eliminate pickup, and connecting wires should be a short as possible to eliminate on–board crosstalk.

The second effect is ground bounce. The ground voltage changes relative to the power supply voltage, and each changes relative to true earth ground, due to currents flowing in the wires between the power supply and the devices on the board. For a digital board, ground bounce may be as high as 0.5 V or more. This 0.5 V change in the bias across the photodiode will generate a

[1] "Cleanliness is next to Godliness" in optical systems.

change in current which may be comparable to the photocurrent, producing spurious data.

Figure 5-12 shows how power and grounds should be routed to minimize ground bounce. Note that analog circuits, if used, may share the same quiet ground pin. For the highest sensitivity and lowest noise performance, analog and digital grounds should be routed separately all the way back to the power supply. However, it may be possible to join the two grounds at the edge of the board.

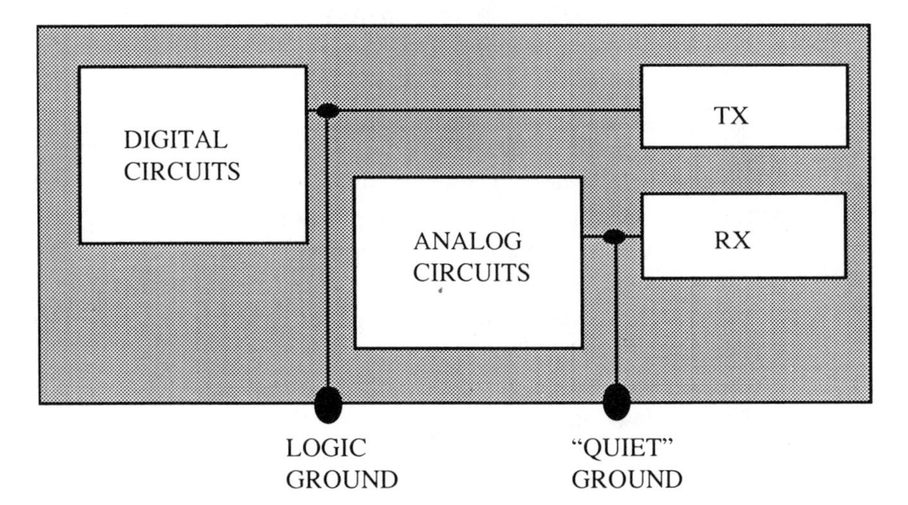

Figure 5-12: Board Layout.

5.8: Summary

A receiver consists of a photodiode, a pre–amp, and other support circuits. The key features of a receiver are the sensitivity and bandwidth.

A photodiode may be a simple PIN structure or an APD with internal gain. A PIN photodiode is very reliable and can be operated from standard board power supplies. An APD is also reliable, but requires a stable high-voltage supply. The APD first breakdown occurs between 25 and 200 V, depending on the device.

The responsivity of a photodiode depends on the material. Silicon is a good material for the short–wavelength window, and several III–V compounds are suitable for the long–wavelength window.

The bandwidth of a receiver is primarily determined by the RC time constant of the amplifier input node and the post–amp filter or equalizer. Bandwidth is sensitive to layout parasitics, particularly wiring capacitance at the pre–amp input node.

Noise originates in the photodiode, the resistors, and the pre–amp transistors. Noise increases with temperature and with APD gain. Noise can be reduced through careful board layout.

5.9: Exercises

1. A photodiode is made from GaAs. What is the cutoff wavelength?
2. An InGaAsP PIN photodiode has a responsivity of 0.64 μA/μW. (a) What is the photocurrent if the input optical power is 50 nW? (b) If it is 1 μW?
3. An APD has a responsivity of 0.55 A/W at 5 V of bias and 4.2 A/W at 30 V reverse bias. What is the gain at 30 V?
4. An APD has a responsivity of 0.55 A/W at 5 V of bias and has a gain of 7 at 30 V reverse bias. What is the responsivity at 30 V?
5. An APD has a responsivity of 0.55 A/W at 5 V of bias and has a gain of 7 dB at 30 V reverse bias. What are the responsivity and gain at 30 V?
6. A PIN photodiode has a dark current of 1.7 nA and a responsivity of 0.55 A/W. What is the necessary input optical power if the photocurrent is to exceed the dark current by 10 dB?
7. An APD photodiode has a dark current of I_{db} = 2.4 nA and I_{ds} = 0.9 nA, a gain of 7, and a low–bias responsivity of 0.55 A/W. What is the necessary input optical power if the photocurrent is to exceed the dark current by 10 dB?
8. An APD photodiode has a low–bias dark current of I_{db} = 2.4 nA and I_{ds} = 0.9 nA, a gain of 7 dB, and a low–bias responsivity of 0.55 A/W. What is the necessary input optical power if the photocurrent is to exceed the dark current by 10 dB? Use $x = 1.5$.
9. A receiver has a measured sensitivity of 45 nW. Express this in dBm.
10. A receiver has a measured sensitivity of –35 dBm. Express this in nW.
11. A receiver is to be selected that will accept 4B/5B encoded data at 100 Mbit/sec (prior to encoding). What is the fundamental frequency of the encoded signal?
12. A receiver is to be selected that will accept Manchester–encoded data at 10 Mbit/sec (prior to encoding). What is the fundamental frequency of the encoded signal?
13. A high–impedance receiver has a bias resistor of 20 kΩ and a junction capacitance of 5 pF. The photodiode is DC coupled to a pre–amp with an input impedance of 100 MΩ and an input capacitance of 7 pF. Calculate the receiver bandwidth.

14. A high–impedance receiver has a bias resistor of 10 kΩ and a junction capacitance of 3 pF. The photodiode is DC coupled to a pre–amp with an input impedance of 100 MΩ and an input capacitance of 2 pF. Calculate the receiver bandwidth.
15. A transimpedance receiver has a bias resistor of 20 kΩ and a junction capacitance of 5 pF. The photodiode is DC coupled to a pre–amp with an input impedance of 5 MΩ, an input capacitance of 7 pF, and a feedback resistance of 60 kΩ. Calculate the receiver bandwidth.
16. A transimpedance receiver has a bias resistor of 10 kΩ and a junction capacitance of 3 pF. The photodiode is DC coupled to a pre–amp with an input impedance of 5 MΩ, an input capacitance of 7 pF, and a feedback resistance of 60 kΩ. Calculate the receiver bandwidth.
17. A high–impedance receiver is to be selected. The bias resistor is 10 kΩ and the photodiode has a junction capacitance of 3 pF. The amplifier has a minimum input impedance of 10 MΩ. If the desired receiver bandwidth is 1 MHz, what is the maximum acceptable input capacitance?
18. A high–impedance receiver is to be selected. The bias resistor is 10 kΩ and the photodiode has a junction capacitance of 3 pF. The amplifier has a minimum input impedance of 10 MΩ. If the desired receiver bandwidth is 130 MHz, show that there is no amplifier with an acceptable input capacitance.
19. A photodiode has a maximum junction capacitance of 5 pF. The pre–amp has an input resistance of 10 MΩ and an input capacitance of 5 pF. If the desired receiver bandwidth is 50 MHz, what is the maximum acceptable bias resistor value?
20. A PIN photodiode has a junction capacitance of 3.4 pF and a leakage current of 1.7 nA. It is connected to a 10–kΩ bias resistor and a high–impedance pre–amp having an input impedance of 10 MΩ, an input capacitance of 7 pF, and noise spectral densities of e_n= 50 nV/Hz$^{1/2}$ and i_n = 0.8 nA/Hz$^{1/2}$.
    (a) Determine the bandwidth.
    (b) Compare the photodiode shot noise, the thermal noise, and the amplifier noise.
21. A PIN photodiode has a junction capacitance of 3.4 pF and a leakage current of 1.7 nA. It is connected to a 10–kΩ bias resistor, a 20–kΩ feedback resistor, and a pre–amp having an input impedance of 10 MΩ, an input capacitance of 7 pF, and noise spectral densities of e_n=50 nV/Hz$^{1/2}$ and i_n=0.8 nA/Hz$^{1/2}$.
    (a) Determine the bandwidth.

(b) Compare the photodiode shot noise, the thermal noise, and the amplifier noise.

(c) Find the NEP.

22. An APD photodiode has a junction capacitance of 34 pF, leakage current of I_{db} = 3 nA and I_{ds} = 2 nA, and x = 0.5. It is connected to a 10–kΩ bias resistor, and a high–impedance pre–amp having negligible noise, an input impedance of 10 MΩ, and an input capacitance of 7 pF. The link has a modulation index of 0.5 and an M = 1 photocurrent of 100 nA. Plot the SNR in dB for gains M = 1 to M = 50. Identify the gain which produces the maximum SNR. (Hint: read the graph, as the maximum may occur between calculated points.)
23. An input optical signal of 76 nW is incident on a PIN photodiode having responsivity of 0.49 A/W. The RMS shot noise is 2 nA, the RMS thermal noise is 3 nA, and the amplifier RMS noise is 0.7 nA. The link has a modulation index of 0.5. Calculate the SNR.
24. An input optical signal of 800 nW is incident on a PIN photodiode having responsivity of 0.52 A/W. The RMS shot noise is 12 nA, the RMS thermal noise is 25 nA, and the amplifier's RMS noise is 12 nA. The link has a modulation index of 0.5. Calculate the SNR.
25. Discuss the advantages and disadvantages of a PIN compared to an APD photodiode.
26. Discuss the advantages and disadvantages of a high–impedance pre– amp compared to a transimpedance pre–amp.
27. Why is a low–impedance pre–amp never used for a fiber optic receiver?

Chapter 6 Optical Fiber Characteristics

The optical fiber carries the optical signal between the transmitter and receiver. Although free space transmission can be used, fiber offers a number of advantages. Fiber carries light around corners, so that fiber optic systems are not limited to line–of–sight transmission, and alignment of transmitter and receiver is not required. Furthermore, fiber attenuation is independent of environmental conditions such as rain and fog.

Fibers for communication systems must meet a number of requirements. The fiber must have low attenuation and high bandwidth, properties easily achieved with silica glass fibers. The fiber must have good optical coupling of light from the transmitter and into the receiver. The fiber must be easily bent (to go around corners), and must have good pull strength so that it can be pulled into place. The fiber must also have high reliability and low cost.

Plastic fiber is used for short links. It has poor performance compared to glass fiber, having both high attenuation and low bandwidth. Plastic fiber is used in areas such as factory automation, where the main concerns are EMI and spark and shock hazards. Plastic–clad silica (PCS) fiber offers a performance between that of solid glass and of solid plastic fibers.

Single strands of glass fiber do not have the pull strength needed for many installations. Individual fibers are also prone to breakage under bending stress, particularly if the surface has been scratched. For this reason most fibers are coated with a protective plastic buffer layer and then placed into a cable. A cable may carry one or several fibers. Common fiber counts include 1 (simplex), 2 (duplex), 6, 10, 12, and 144.

A hybrid cable contains both wire and fiber lines. Hybrid cables are used to provide an office room with multiple connections, such as twisted pair for telephone and fiber for the company computer network. Hybrid fibers are also put in place where current end electronics connect to wire systems, but future upgrades to fiber connections are planned.

6.1: Fiber Profiles

A fiber is constructed of a core surrounded by a cladding. The core and cladding are both made from glass, with only a small difference in doping. To achieve guided propagation, the core index of refraction must be higher than the cladding index. The cross–section of a fiber is shown in Figure 6-1. The core and cladding are ideally concentric circles. In reality, core eccentricity and non–concentric structure will degrade fiber quality slightly. The outer diameter is normally 125 μm, to fit standard fiber optic connectors, although other outer diameters are available.

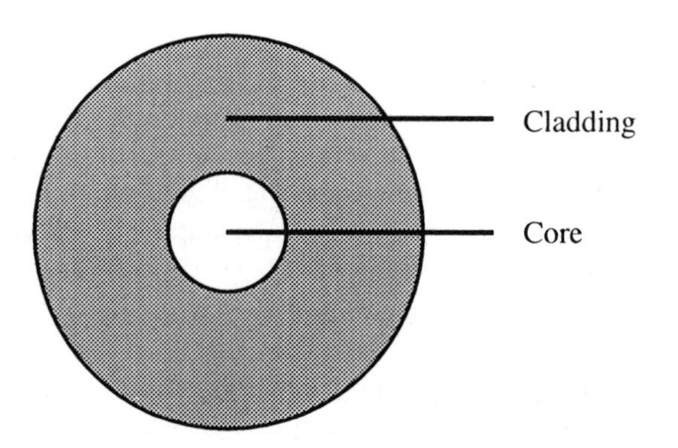

Figure 6-1: Fiber Cross–section.

The fiber is characterized by its optical characteristics, which are determined by the glass doping (intentional impurity concentration). The symbol n_1 is used for the core index, and n_2 for the cladding index. The index difference, Δ, and the numerical aperture (NA) are related to the core and cladding indexes:

$$\Delta = (n_1 - n_2) / n_1 = NA^2 / (2 n_1^2)$$

$$NA = (n_1^2 - n_2^2)^{1/2} = n_1 (2 \Delta)^{1/2}$$

For typical fibers, Δ is on the order of 0.01 (usually between 0.005 and 0.02). Usually only n_1 and NA are given on data sheets.

EXAMPLE:

A fiber has a core index of 1.448 and a numerical aperture of 0.21. Find the index difference and the cladding index.

a) $\Delta = NA^2 / (2\,n_1^{\;2}) = (0.21)^2 / (2 \times 1.448^2) = 0.0105$

b) $n_2 = n_1(1 - \Delta) = (1.448)\,(1 - 0.0105) = 1.433$

There are three fiber profiles in commercial use. The oldest, and lowest cost, is the step–index fiber. The second oldest, and most expensive, is the graded–index fiber. The newest is the single–mode fiber. A fiber is characterized by its profile and by the core and cladding diameters. For example, a particular fiber may be described as "STEP 62.5/125", which indicates that is has a step–index profile, a core diameter of 62.5 μm, and a cladding (outer) diameter of 125 μm. Figure 6-2 shows the index profile for the three fiber types.

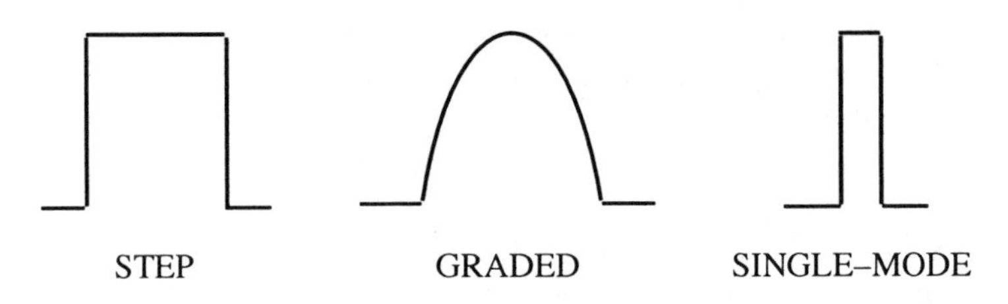

Figure 6-2: Index Profiles.

The step–index fiber is made from a higher index rod of glass surrounded by a lower index glass coating. As the rod is heated and drawn, the glass diameter becomes smaller. After drawing, the core typically has a diameter of 50–80 μm, and the cladding has a diameter of 125 μm. The standard sizes are 50/125 and 62.5/125. Some older systems also use 80/125 and 100/140, although these are not being used for newer systems. Experience has shown that the wider core diameter does not significantly increase light coupling efficiency (since that is determined by the core and cladding indexes of refraction), while the bandwidth capability is decreased.

The graded–index fiber has a core made from many layers of glass. The index of refraction of each layer is slightly lower than that of the next inner layer. A graded–index fiber has lower coupling efficiency and higher bandwidth than the step–index fiber. It is available in 50/125 and 62.5/125 sizes. The 50/125 fiber has been optimized for long–haul applications, and has a smaller NA and higher bandwidth. 50/125 fiber costs approximately 25% less than 62.5/125 fiber, which is optimized for LAN applications.

The grading of the index profile across the fiber diameter has a strong influence on both coupling efficiency and on pulse spreading along the fiber length. The most common way to describe the profile is to approximate the index of refraction as a function of radius using a parabolic function:

$$n(r) = n_1 \sqrt{1 - 2\Delta \left(\frac{r}{a}\right)^{\alpha}}$$

$$NA(r) = NA(0)\sqrt{1 - \left(\frac{r}{a}\right)^{\alpha}}$$

where a is the fiber radius (d/2), and α is the index profile grading characteristic.[1]

Both step–index (SI) and graded–index (GI) fibers carry thousands of optical modes, and both are called multimode (MM) fibers. For a graded–index fiber, the number of modes decreases as α decreases, so that a parabolic graded–index ($\alpha = 2$) fiber carries approximately half the modes of a step-index ($\alpha = \infty$) fiber. The actual number of modes in a graded–index fiber can be calculated from:

$$M_{GI} = M_{STEP} \, [\alpha / (\alpha + 2)]$$

Single–mode (SM) fiber carries only the lowest order mode. The single–mode fiber is actually a step–index fiber which has been pulled to a very thin core diameter, typically 5–10 μm. The size of the core is more typically given in terms of the "mode field diameter", or the diameter of the optical power distribution, which is significantly wider than the core diameter. Single–mode fibers have the lowest coupling efficiency and highest bandwidth of the three types of fibers.

6.2: Waveguide Properties

The solutions to Maxwell's equations for a cylindrical fiber profile are given in Appendix E for a step–index fiber. Solutions for graded–index fibers can be found only when the exact grading profile is known, which is generally not true. The characteristics for actual graded–index fibers are not identical with the derived characteristics for parabolic fibers, since n(r) may not follow the parabolic mathematical formula.

---

[1] Note that this α is dimensionless, and different from the attenuation factor α.

The number of solutions is related to the V–parameter (also referred to as the V number):

$$V = \left(\frac{\pi d}{\lambda}\right) * \mathrm{NA}$$

where d is the fiber diameter, λ is the optical carrier wavelength, and NA is the numerical aperture.

The number of modes M (for $V > 20$) is approximately:

$$M = V^2 / 2$$

The fraction of the optical power traveling in the cladding is:

$$P_{clad} / P_{tot} = \frac{4}{3\sqrt{M}}$$

For V smaller than 20, these approximations are not valid. For example, for single–mode fiber (M = 1), the equation predicts that 4/3 of the total power is on the cladding, which is clearly impossible. For small V, the number of modes can be derived from Maxwell's equations and the Bessel function solutions.

EXAMPLE:

An optical fiber with a core diameter of 50 μm and a numerical aperture of 0.25 carries light with a wavelength of 850 nm. Determine the number of modes and the fraction of the total power that travels in the cladding.

$V = (2 * \pi * 25\ \mu m / 0.82\ \mu m) * 0.25 = 47.9$
Since $V > 20$, the large–V approximation can be used.
$M = V^2 / 2 = 1146$ modes

$$P_{clad} / P_{tot} = \frac{4}{3\sqrt{1146}} = 0.039 = 3.9\%$$

The cutoff condition for single–mode operation is $V = 2.405$. For $V < 2.405$, there is exactly one mode that can travel in the fiber. A fiber having a sufficiently narrow core will be a single–mode fiber for some range of wavelengths. Single–mode fibers provide the highest fiber bandwidth for communications applications.

EXAMPLE:

A fiber has an effective core diameter of 8 μm, a core index of 1.446, and a cladding index of 1.440. What is the wavelength range for which this is a single–mode fiber?

a) The cutoff condition for single–mode operation is V = 2.405.

$$V = \left(\frac{\pi d}{\lambda}\right) * NA \quad = [(2\pi)(4\mu m) / \lambda]\,(1.446^2 - 1.440^2)^{0.5} = 2.405$$

$$\lambda = [(2\pi)(4\ \mu m)(1.446^2 - 1.440^2)^{0.5}] / (2.405) = 1.375\ \mu m$$

Since V decreases as λ increases, this fiber will be single–mode for all wavelengths greater than 1.375 μm or 1375 nm. For example, this fiber will not exhibit single–mode operation in the 1300–nm window, but will in the 1550–nm window.

6.3: Optical Interface

Light enters the fiber from a transmitter. As was seen in Chapter 3, the entering light will be coupled into the fiber only if its entrance angle is less than $\theta_{o,ext}$. Light outside the entrance angle may couple into the fiber, but will suffer high attenuation. The amount of optical power observed along a link will follow a characteristic curve similar to Figure 6-3. The equilibrium length, L_e, is the length at which the loosely bound (leaky) modes have attenuated out and ceased to be significant. For typical silica glass fibers, this length is on the order of a few tens of meters.

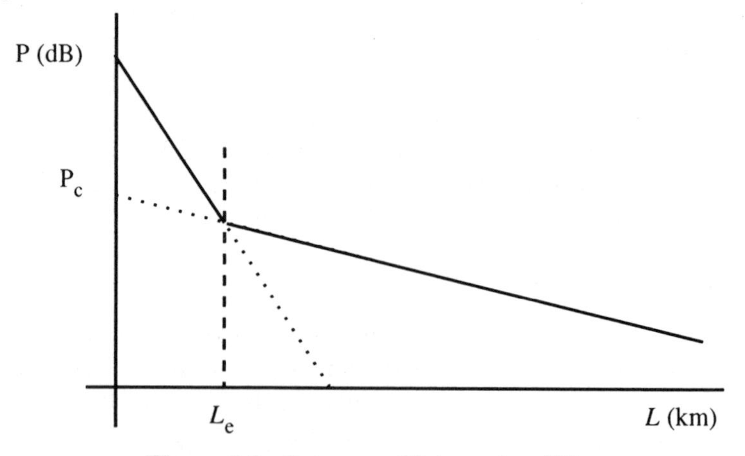

Figure 6-3: Power vs. Distance in a Fiber.

The equilibrium power coupled into the fiber is P_c. For short links (less than L_e) a greater amount of power is present. The amount of power coupled will depend on a number of factors, including the spatial distribution of light incident on the fiber and the fiber profile.

Optical coupling efficiency is normally calculated using the numerical aperture (NA). For an optical source concentric with the fiber and a flat fiber end face, the coupling factor for a surface–emitting LED is given by:

$$P_c = P_s \, (\mathrm{NA})^2 \;\; \mathrm{MIN}\,[\,1, (a\,/\,r_s)^2]$$

where a is the fiber radius, P_c is the power coupled into the fiber from an LED whose output power into air is P_s, and r_s is the LED radius. The coupled power is independent of the relative size of the fiber core and the LED when the LED is smaller than the fiber core.

EXAMPLE:

An LED of diameter 50 μm puts out 0.2 mW. How much power is coupled into a 50/125 step–index fiber whose NA is 0.25?

$$P_c = (0.2\text{ mW})\,(0.25)^2 \;\; \mathrm{MIN}\,[1, (25\,/\,25)^2]$$
$$P_c = (0.2\text{ mW})\,(0.0625)\,(1) = 0.0125\text{ mW} = 12.5\ \mu\text{W}$$

As can be seen from this example, only a small fraction of the optical power is actually coupled into the fiber. Fiber optics is rarely used for power transmission, except for illumination of enclosed areas (e. g. inside engine cylinders). Fiber optics is used predominantly for communications, where the only requirement for the output power is that it be adequate to drive the receiver.

Light from an edge–emitting LED will be more efficiently coupled into a fiber than that from a surface–emitting LED. To calculate exactly how much light will couple into a given fiber, the LED's output spatial distribution must be known (i. e. how much of the optical output is inside the external acceptance angle). The coupling efficiency can be estimated using the surface LED formula if no coupling efficiency specifications are given.

EXAMPLE:

An LED of diameter 50 μm puts out 0.2 mW. How much power is coupled into a 5–μm diameter single–mode fiber whose NA is 0.15?

$$P_c = (0.2\text{ mW})\,(0.15)^2 \;\; \mathrm{MIN}\,[1, (2.5\,/\,25)^2]$$
$$P_c = (0.2\text{ mW})\,(0.0225)\,(0.01) = 45 \times 10^{-6}\text{ mW} = 45\text{ nW}$$

A single–mode fiber is characterized by a small core diameter and NA. This makes it very difficult to couple light from an LED into the fiber. A single–mode fiber is usually used with a laser diode, although edge–emitting LEDs can sometimes be used. For laser diodes, as with edge–emitting LEDs, the optical coupling will depend on the diode's output spatial distribution. Because the laser output has an elliptical spatial distribution, the coupling efficiency will be (roughly) proportional to NA, rather than NA^2.

The coupling between a laser diode and a single–mode (or multimode) fiber is more efficient than for an LED. The spatial spread of light from the laser is narrow, and much of the optical power is inside the external acceptance angle. For many laser diodes, the coupled power is 30-50% of the source power. The NA effect for typical single–mode fibers is included in this rule–of–thumb estimate.

EXAMPLE:

A laser of dimensions 0.5 x 5 μm puts out 0.2 mW. How much power is coupled into a 5 μm single–mode fiber whose NA is 0.15? Assume maximum rule–of–thumb coupling efficiency.

The laser is smaller than the fiber. The effect of the NA is incorporated in the rule–of–thumb estimate.

P_c = (0.2 mW) (50%) = 0.1 mW

A comparison of coupling efficiency for LEDs and lasers with single–mode fibers shows that far more power is coupled with a laser. In the examples above, the laser coupled 0.1 mW, compared to only 45 nW for the LED, a difference of 33 dB.

Coupling efficiency is often expressed in dB. For example, laser coupling efficiency is usually between 3 and 5 dB.[1] For an LED, coupling efficiency is much lower, typically around 11-16 dB, depending on the NA of the fiber.

6.4: Attenuation

Once the power has been coupled into the fiber, the optical signal will interact with the fiber. There are two major interaction effects: attenuation and dispersion. Attenuation decreases the optical power, and dispersion decreases the available bandwidth. Both effects degrade link performance; the longer the link, the greater the degradation.

---

[1] The industry convention is to use a *positive* dB to indicate *loss*.

Attenuation is caused by two physical effects: absorption and scattering. Absorption removes the photons in interactions with atoms and molecules, while scattering redirects the light out of the core.

Absorption occurs when the energy of the photon is equal to a difference between two electronic energies. For example, there are energy levels corresponding to molecular energy states of silica (SiO_2) molecules and electron energy levels in the oxygen, silicon, and impurity atoms. Absorption causes severe attenuation in the UV region,[1] and is a major effect in silica glass for wavelengths above 1600 nm.

Scattering losses occur when the photons see a variation in the core's index of refraction. Photons can be scattered out of the fiber from small density variations in the glass. Density variations may be caused by incomplete molecular bonds, or be variations in molecular spacing. Scattering is significant for wavelengths below 1000 nm.

A major cause of absorption is the presence of OH^- radicals, which results from the presence of water (H_2O). OH^- enters the fiber through either a chemical reaction (fiber manufacturing) by–product, or as humidity in the operating environment. The main OH^- absorption peak is at 1400 nm, with a secondary peak at 950 nm. Humidity reduction during manufacturing brought fiber attenuation from 1000 dB/km down to 20 dB/km in the 800–nm window in 1970. Because it is impossible to provide 0% humidity over the entire life of a fiber, optical communications systems are designed near 1300 nm or 1550 nm, but not near 1400 nm.

The attenuation for a particular fiber may be specified in one or more windows. Generally, a fiber should be used only in those windows for which specifications are given. Attenuation may vary as much as 1 dB/km from one batch to another. Designs should based on the manufacturer's maximum attenuation for that fiber (at the wavelength of interest), rather than on attenuation for any one batch.

The attenuation will vary with the quality of the glass material, impurity types and concentration, optical wavelength, and fiber core diameter. The best attenuation is achieved with the most uniform glass material and the lowest impurity concentration. The lowest attenuation occurs for wavelengths near 1550 nm. The attenuation in the 1550–nm window can be as low as 0.2 dB/km, compared to around 0.5 dB/km in the 1300–nm window. The 850–nm window has higher loss, typically above 3 dB/km. Attenuation increases with

---

[1] Which is why you do not suntan inside the house, no matter how clean the windows.

increasing fiber diameter, since more of the power distribution is near the cladding outer surface (e. g. the cladding is only 62.5 μm thick for 62.5/125 fiber, compared to 75 μm for 50/125 fiber). A graded–index fiber will have a lower attenuation than a step–index fiber, since power is concentrated closer to the center of the fiber.

Commercial fiber comes in a number of grades, depending on profile, the attenuation, and the fiber bandwidth. Typical values of attenuation are given in Table 6-1. Fibers with significantly lower attenuation have been made in the laboratory (e. g. below 2 dB/km at 800 nm); values in the table are typical of commercially available fibers. Figure 6-4 shows the variation of attenuation with wavelength.

Table 6-1: Attenuation for Typical Fibers

λ	Fiber type	Size	Attenuation (dB/km)
800	Step	62.5/125	5.0
850	Step	62.5/125	4.0
	Graded	62.5/125	3.3
	Graded	50/125	2.7
	Single–mode	x / 125	
1300	Graded	62.5/125	0.9
	Graded	50/125	0.7
	Single–mode	x / 125	0.5
1550	Single–mode	x / 125	0.2

Attenuation leads to a loss of power along the fiber. The output power is significantly less than the coupled power. The output power P_o at the end of a fiber link depends on the coupled power (P_c), the attenuation per unit length (α), and length (L):

$$P_o = P_c \; 10^{(-\alpha L / 10)} \qquad \text{(mW)}$$

$$P_o = P_c - \alpha L \qquad \text{(dB)}$$

Note that the two equations describe the same power levels. It is important to know what power units are in use. Watts (or mW, μW, nW) should never be mixed with dBW (or dBm, dBu, dBn) within an equation, except when converting between them.[1]

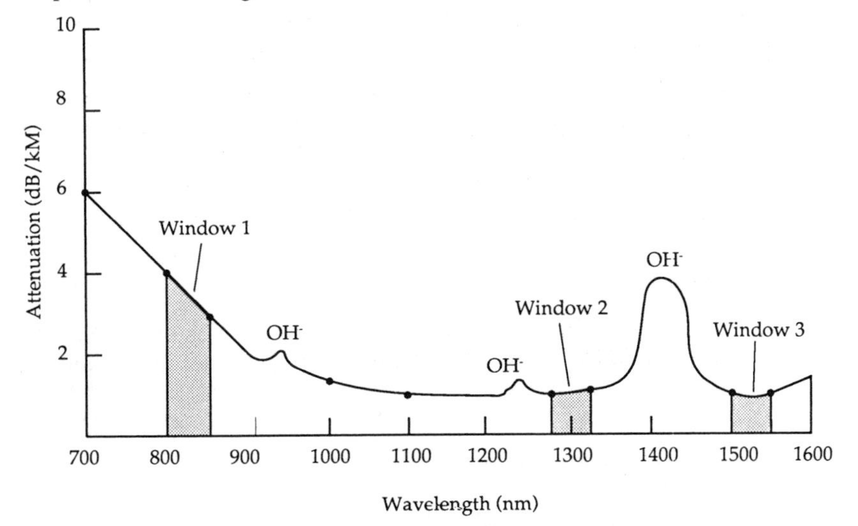

Figure 6-4: Attenuation vs. Wavelength.

Courtesy of The Light Brigade Inc.

EXAMPLE:

A fiber has a coupled power of 0.16 mW, attenuation of 6 dB/km, and a length of 2 km. What is the output power?

Since the power is given in mW, use the first form of the equation:

$P_o = P_c\ 10^{(-\alpha L / 10)} = (0.16\ \text{mW})\ 10^{[-(6\ \text{dB/km})(2\ \text{km})\ /\ 10]}$

$P_o = 10\ \mu\text{W} = -20\ \text{dBm}$

EXAMPLE:

A fiber has a coupled power of –8 dBm, an attenuation of 6 dB/km, and a length of 2 km. What is the output power?

Since the power is given in dBm, use the second form of the equation:

$P_o = P_c - \alpha L = (-8.0\ \text{dBm}) - (6\ \text{dB/km})(2\ \text{km})$

$P_o = -20\ \text{dBm} = 10\ \mu\text{W}$

[1] P (dBm) = 10 log (P in mW) P (mW) = $10^{P\,(\text{dBm})/10}$

Both forms of this equation produce the same calculated output power. The first form is used when the input and output power are both in mW. The second form is used if either of the powers is in dBm.

EXAMPLE:

A fiber has a coupled power of –8 dBm, an attenuation of 6 dB/km, and a required output power of –30 dBm. What is the maximum fiber length?

$$L = (P_o - P_c) / \alpha = [(-8) - (-30)] / (6 \text{ dB/km}) = 3.67 \text{ km}$$

EXAMPLE:

A fiber has a coupled power of –8 dBm, an attenuation of 1 dB/km, and a required output power of –30 dBm. What is the maximum fiber length?

$$L = (P_o - P_c) / \alpha = [(-8) - (-30)] / (1 \text{ dB/km}) = 22 \text{ km}$$

The maximum transmission distance due to attenuation is inversely proportional to the fiber attenuation α. For a link shorter than 2 km, it may be possible to operate in the 850–nm window. For metropolitan and wide area networks, spanning up to 10 and 50 km, respectively, the preferred operating wavelength is 1300 nm. For long–haul networks, use of the 1550–nm window allows link spans of 40 miles or more.

6.5: Bending Losses

Bending losses occur when the fiber is bent. Macrobends, visible from outside the cable, are most commonly encountered in routing the fiber around corners. Macrobends may extend through a large arc, even a quarter of a circle or even a full loop. Microbends, which cannot be seen from outside the cable, are caused by the cabling process and by compressive forces on the fiber. A microbend cannot extend through more than a small arc. (A bend radius of less than 1 cm through a visible arc will almost certainly break the fiber.) Figure 6-5 shows the two types of bends.

The amount of macrobending loss depends on the fiber profile and the number of leaky modes present. A derivation by Gloge is often used to predict the bending loss.[1] For a multimode fiber, the power transmission around a

[1] D. Gloge, "Bending Loss in Multimode Fibers with Graded and Ungraded Core Index", *Applied Optics* **11** (Nov 1992) 2506–2512.

bend of radius R is approximately:

$$\frac{P_2}{P_1} = 1 - \left(\frac{\alpha + 2}{2\alpha\Delta}\right)\left[\frac{d}{R} + \left(\frac{3\lambda}{4\pi n_2 R}\right)^{\frac{2}{3}}\right]$$

where P_1 and P_2 are in mW, d is the fiber diameter, α is the profile grading factor, Δ is the index difference, n_2 is the cladding index, and λ is the optical wavelength. This formula applies only when R is much larger than d, which is true for macrobends.

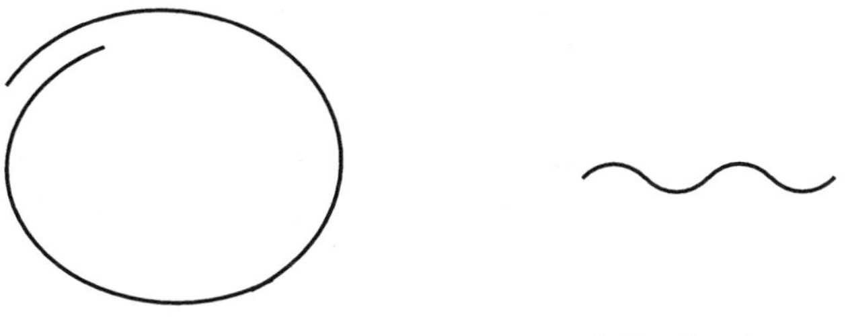

Macrobend Microbend

Figure 6-5: Macrobends and Microbends.

The macrobending loss increases as the radius R gets smaller, as the fiber core size becomes smaller, and as the wavelength increases. Macrobending losses are higher for step–index than for graded–index fibers.

The bending loss for a reel of fiber is frequently treated as a bent–fiber attenuation. For a number of turns, a large bending loss occurs in the first turn, and a smaller bending loss will occur for each additional turn. Additional loss from the inherent fiber attenuation (αL) occurs in addition to the bending loss.

EXAMPLE:

A 62.5/125 step–index fiber has a core index of 1.448 and a numerical aperture of 0.21, and carries 1300 light around a 2–cm turn. What fraction of the power is lost in the turn?

First calculate the necessary values: $\Delta = 0.0105$ and $n_2 = 1.433$.
For a step–index fiber, $\alpha = \infty$, and $(\alpha + 2)/\alpha = 1$.
$1 \text{ cm} = 10^4 \ \mu\text{m}$

$P_2 / P_1 = 1 - (1 / 2\Delta)$
$\quad \times \ [(62.5 \ \mu\text{m} / 2 \text{ cm}) + \{(3)(1.30 \ \mu\text{m}) / (4\pi)(1.433)(2 \text{ cm})\}^{2/3}]$
$P_2 / P_1 = 1 - 0.17 = 0.83$
17% of the power is lost. This result can also be expressed in dB:
$P_{lost} = 10 \log (P_2 / P_1) = 0.82 \text{ dB}$

EXAMPLE:

A 62.5/125 graded–index fiber has a core index of 1.448 and a cladding index of 1.433, and carries 1300 light around a 2–cm right angle turn. What fraction of the power is lost in the turn?

For a graded–index fiber, $\alpha = 2$, and $(\alpha + 2)/\alpha = 2$.

$$P_2 / P_1 = 1 - (2 / 2\Delta) \times [(62.5\ \mu\text{m} / 2\ \text{cm}) + \{(3)\ (1.3\ \mu\text{m}) / (4\ \pi)\ (1.433)\ (2\ \text{cm})\}^{2/3}]$$

$$P_2 / P_1 = 0.66$$

34%, or 1.8 dB, of power is lost.

As seen in the example, graded–index fibers have higher bending losses than step–index fibers. Bending losses are highest for single–mode fibers, particularly at wavelengths above the cutoff wavelength. For example, a fiber designed with a 900–nm cutoff wavelength, optimized for 1300–nm single-mode operation, will have highest bending losses in the 1550–nm window. Since the actual bending losses depend significantly on the fiber profile, bending loss information must be obtained from the vendor or through direct measurements.

In addition to causing power loss, bending causes mode stripping and mixing. Macrobends result in mode stripping, where the outermost modes refract out of the fiber, as shown in Figure 6-6. After mode stripping, the power distribution in the fiber is more centralized, with less power in the outer core and cladding. A mode–stripped fiber will exhibit lower attenuation.

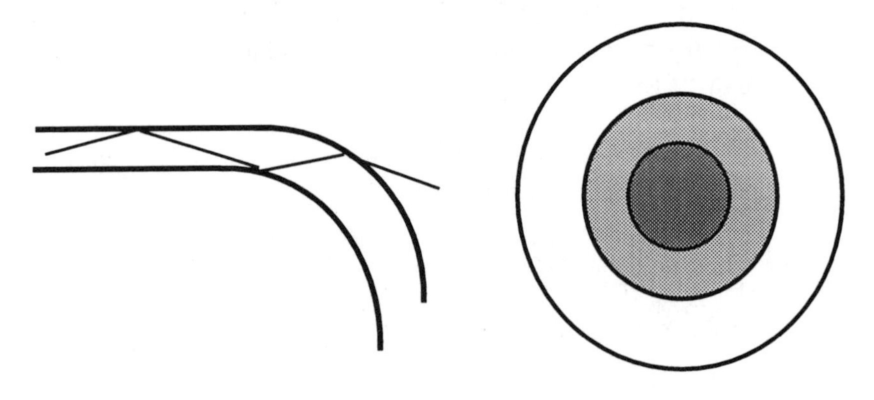

Figure 6-6: Mode Stripping.

Microbends contribute to mode mixing as well as mode stripping. Microbends cause the power in the innermost modes to "mix" with, or couple into, outer modes, as well as causing increased loss of those outer modes. As power is coupled from inner modes to outer modes, fiber attenuation increases.

Macrobends are intentionally introduced into short fibers to simulate the effects of long fibers. Macrobending reduces the equilibrium length to a few centimeters, since all leaky modes are removed at the bends. However, outer modes that contribute to loss in the long fiber may also be stripped out if the bending radius is too tight, leading to underestimating the attenuation of the installed fiber. Many vendors use mode strippers that leave the core about 70% mode filled, meaning that the outermost 30% of the modes are removed by stripping. Mode stripping, when used for design performance prediction, must be used with care, and comparisons to actual installations should be used to improve future design estimates.

6.6: Connection Losses

Additional losses occur whenever there is a connection point between two fibers or between a fiber and LED or photodiode. The loss may be low, as when two fibers are fused together; or it may be higher, as when an LED is coupled to a fiber. This section covers losses due to mismatches in optical properties and mechanical alignment.

A mismatch in optical properties causes a Fresnel reflection loss. When light enters from air (n = 1), there is an index of refraction mismatch with the fiber. This causes the fiber endface to act like a mirror, reflecting some of the light. The power transmission though a partially reflective boundary is given by:

$$\frac{P_2}{P_1} = 1 - \left(\frac{n_1 - n}{n_1 + n}\right)^2$$

where n_1 and n are the two indexes of refraction. When $n_1 = n$, there is no reflection, and $P_2 = P_1$. The larger the index mismatch, the more power will be reflected. If the endface of the fiber is not perpendicular to the fiber axis, the reflection induced loss will be greater.[1]

When index mismatch leads to reflections, an index matching material may be inserted into the air gap. The minimum reflection loss will occur when

[1] For laser–based systems operating at high data rates, non–perpendicular endfaces are used to prevent reflected light from entering back into the laser (which would treat it as a signal and re–amplify and re–transmit it).

the index matching material has an index of refraction:

$$n_m = \sqrt{n_1\, n}$$

For example, to reduce reflections between a core index of 1.448 and air (n = 1), the ideal index matching material has $n_m = 1.203$.

A mismatch in alignment of the two fibers will also cause a loss. Mechanical effects at connection points include physical alignment and fiber size mismatches. Physical alignment involves all three dimensions: core offset, core separation along the common axis, and core directional angle from the first axis. Fiber size mismatch includes the core diameter, the core profile, and the core eccentricity.

Figure 6-7 shows the three physical alignment terms. θ is the angle between the two core axes, d is the axial offset,[1] and s is the separation of the two endfaces. Misalignment in each of these three directions leads to additional power transmission losses at the connection point.

Fiber mismatch due to axial offset is caused by the two cores not being concentric. The major cause of offset is the alignment tolerances in splice and connector hardware. Another cause is the slightly non–concentric fiber structure that may occur due to limitations in the manufacturing process. Core eccentricity, or non–circularity, results in a similar effect.

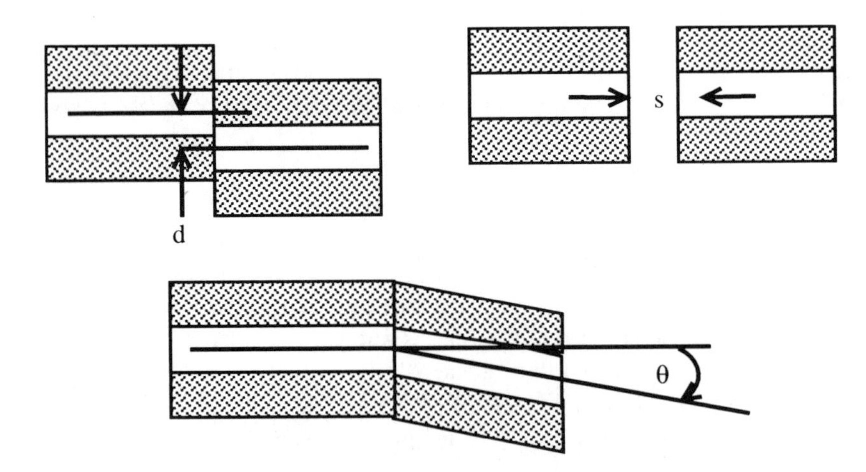

Figure 6-7: Physical Alignment.

[1] Not to be confused with d, the fiber diameter.

For a step–index fiber, the power transmission ratio for the axial offset is given by:

$$P_2 / P_1 = \text{MIN}\,(1, \{ [(2/\pi) \arccos (d/2a)] - [(d / \pi a)\ (1 - \{d/2a\}^2)^{1/2}] \})$$

where d is the axial offset and a is the fiber diameter.

For graded–index fibers having $d/a < 0.4$, as is true in most connections, the axial offset loss can be approximated by:

$$P_2 / P_1 = 1 - 8\, d / (3\, \pi\, a)$$

Endface separation is usually the result of the small distance between two fibers in a connector. Since even the smallest scratch increases optical loss, connectors are designed to keep a small (few micron) space between the two fiber endfaces. For endface separation, the power transfer ratio is:

$$P_2 / P_1 = \text{MIN}\,(1, a / [a + s \tan \theta_{o,ext}])$$

where a is the fiber radius and s is the endface separation.

Angular misalignment results from fibers not being perfectly aligned in a connector or splice. For angular misalignment, the transmission loss equation, given by Thiel and Hawk,[1] is:

$$\frac{P_2}{P_1} = \cos\theta \left\{ \frac{1}{2} - \frac{p}{\pi}\sqrt{1-p^2} - \frac{1}{\pi} \arcsin(p) - q \left[\frac{y}{\pi}\sqrt{1-y^2} + \frac{1}{\pi} \arcsin(y) + \frac{1}{2} \right] \right\}$$

where the parameters p, q, and y are in turn given by:

$$p = \cos\theta_c\, [\, (1 - \cos\theta) / (\sin\theta_c \sin\theta)\,]$$

$$q = \cos^3\theta_c\ / (\cos^2\theta_c - \sin^2\theta)^{3/2}$$

$$y = [\, (\cos^2\theta_c)\, (1 - \cos\theta) - \sin^2\theta]\ / (\sin\theta_c \cos\theta_c \sin\theta)$$

Connection losses also occur due to any difference in physical characteristics between the two fibers. The physical mismatches that can occur include fiber core size, fiber profile, and fiber NA.

When the receiving fiber core is smaller than the emitting core, some of

---

[1] F. Thiel and R. Hawk, "Optical Waveguide Cable Connection", *Applied Optics* **15** (Nov. 1976) 2785–2791.

the power couples into the receiving fiber cladding, where it is lost. The core size mismatch causes a power transmission ratio of:

$$P_2 / P_1 = \text{MIN}\,(1,\ (d_R / d_E)^2)$$

where d_R is the receiving fiber diameter and d_E the emitting fiber diameter.

A profile mismatch occurs when the two fibers have different index profiles. This occurs when a step–index fiber is connected with a graded–index fiber, or when two different grading profiles are present. The core profile mismatch causes a power transfer ratio of:

$$P_2 / P_1 = \text{MIN}\,(1,\ [\,(\alpha_R)(\alpha_E + 2)\,] / [\,(\alpha_E)(\alpha_R + 2)])$$

where α_R and α_E are the grading parameters (not the attenuation!).

A mismatch in numerical aperture may be caused by a difference in core index, in cladding index, or in both. The power transmission through an NA mismatched connection is:

$$\frac{P_2}{P_1} = \text{MIN}\,[\,1, (\frac{NA_R}{NA_E})^2]$$

where NA_E is for the emitting fiber, and NA_R is the for the receiving fiber. When the receiving fiber has the higher NA, there is no loss due to NA mismatch.

EXAMPLE:

Two fibers, each having a core index of 1.445, are coupled with an air gap between them. The first fiber has an NA of 0.25, and the second fiber, 0.20. Find the power transmission ratio.

There are two reflections, one at each air–fiber interface. Each interface has a transmission ratio of:

$$P_2 / P_1 = 1 - [(1.445 - 1)/(1.445 + 1)]^2 = 0.967$$

The receiving fiber has a lower NA, so the NA transmission ratio is:

$$P_2 / P_1 = (0.20 / 0.25)^2 = 0.640$$

The net effect of several losses is calculated by multiplying the transmission ratios:

$$P_{out} / P_{in} = (0.967)\,(0.967)\,(0.640) = 0.60$$

There are a variety of physical connectors available. Each has its advantages and disadvantages. Of greatest interest are the connection loss, the reliability, and the cost. Figure 6-8 shows some typical connectors; some are similar to the BNC connector, while others are screw–on types similar to SMA connectors. The physical outer dimensions of the connector are limited by the need to manually connect them together, while the inner dimensions are limited by the alignment requirements of the fibers.

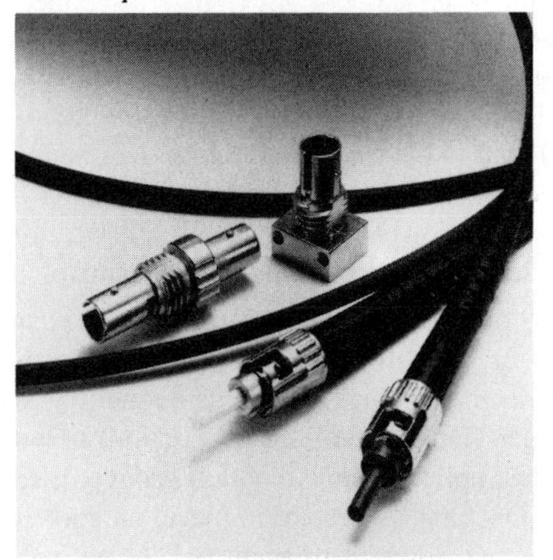

Figure 6-8: Connector Styles.

Photograph courtesy of AMP Inc.

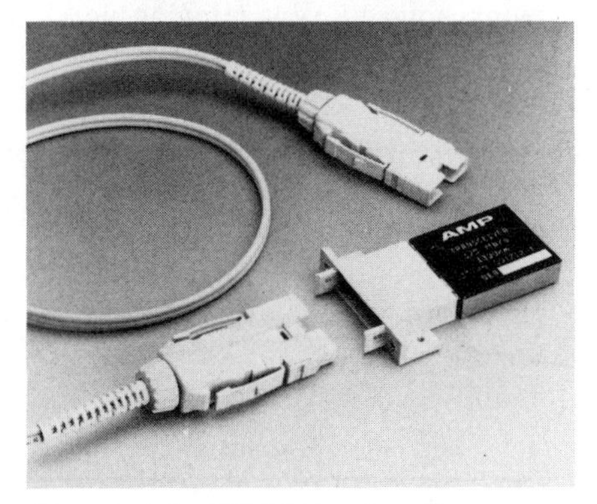

Figure 6-9: FDDI Duplex Connector.

Photograph courtesy of AMP Inc.

Standards for connectors are slowly evolving. While two connectors may be designed to have the same external style connecting the same grade of fiber, they are not interchangeable. Thus connectors from two different vendors will mate physically but may have a higher loss than if the two connectors were from the same vendor. This will become less of a problem as standards evolve. At the present time, the FDDI connector, shown in Figure 6-9, is the best defined connector standard.

6.7: Splices And Connectors

Splices are used to permanently join two fibers. Connectors are used where the connection may be disconnected as needed. Splices are characterized by low loss and pull strengths comparable to that of the fiber. Connectors are designed to have repeatable loss over many connect/disconnect cycles, strength under handling, and moderate loss. Typical splice loss is 0.15–0.2 dB, for both fusion and mechanical splices. Connector losses are typically 0.5–3 dB, depending on the type of connector.

There are two common ways of making splices. Fusion splices are made by actually melting the two fiber endfaces and fusing them together. Fusion splices are as strong as the original fiber. Fusion melting is accomplished with either a small hydrogen flame (in volume production) or with an electric arc (in the field). Mechanical splice fixtures contain grooved structures which guide the fibers together. Once the fibers are in place, an epoxy is used to bond the splice fixture and the two fibers permanently. The epoxy itself may also provide index matching within the enclosure, or an index matching material may be used inside the fixture. Figure 6-10 shows the structure of a typical mechanical splice showing how the center region necks down to guide the fibers together. Today, mechanical splices can be made with losses as low as 0.15 dB, comparable to a fusion splice.

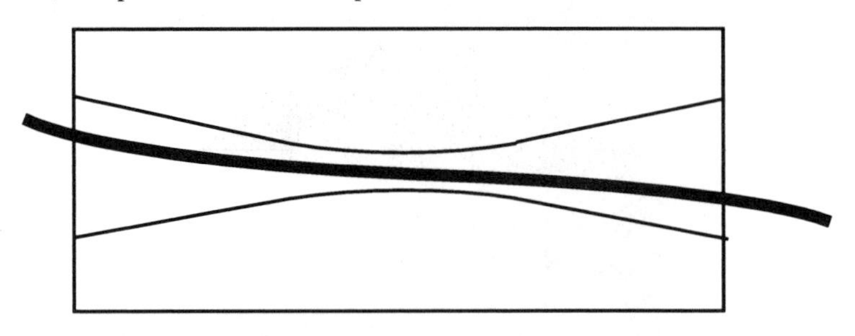

Figure 6-10: Mechanical Splice Scheme.

Fusion splices have the advantage of low cost per splice, since there is no fixture attached permanently to the fiber. Fusion splicing requires a relatively high initial capital cost. For an electrical arc splicer, for example, a field splice system must include not only the splicer, but also the truck to transport it and a generator to produce the electrical arc. Mechanical splices have a low initial capital cost, usually only that of the epoxy cure system (if any), but have a high component cost per splice. Mechanical splices can be "gang–cured", where a group of splices is cured with a UV cure light. Fusion splicing is preferred in assembly environments, while mechanical splicing is used in small volume and LAN applications.

There are three standard types of connectors in use for local area network and long–haul communications, and at least 25 connector types commercially available. Connector types do not interconnect, so jumper cables are required to make connections between different connector types.

Most connectors are designed to polish each fiber endface at the end of a ferrule, and an alignment sleeve butts the ferrules with a fixed endface separation. The outside (finger contact) styles for connectors include screw–on, BNC–mount, bayonet, and push–pull mechanical contacts. Most connector styles, except those designed for plastic fiber, include an alignment ferrule and sleeve designed to guide the two fiber endfaces close to contact, as shown in Figure 6-11.

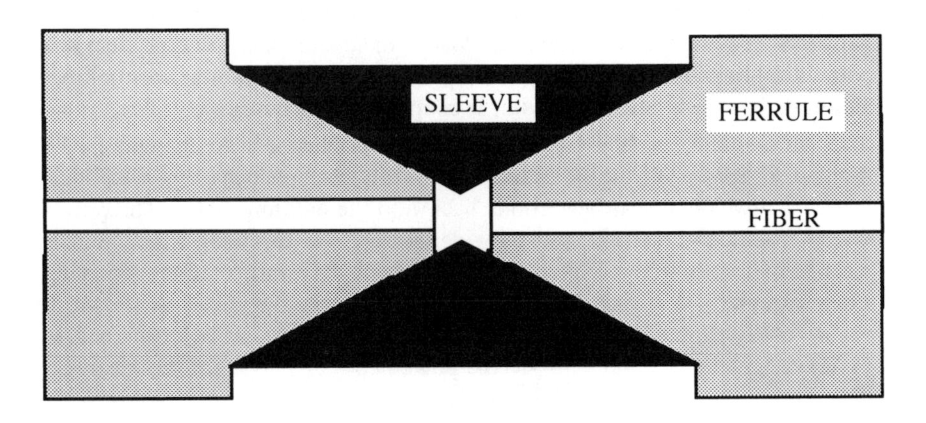

Figure 6-11: Tapered (Biconic) Ferrule and Sleeve.

Expanded beam connectors use small lenses in each half of the connector to expand the emitted beam and focus it into the receiving fiber. Expanded beam connectors are more expensive than the butt connectors, and are used primarily for single–mode connections. Because single–mode fiber has a small diameter, the loss due to even a small axial offset can be very large, and the expanded beam reduces sensitivity to alignment.

Connector loss is usually quoted for connecting two ends of one fiber together, and does not include any losses due to differences in the two fibers being connected. Actual losses in the field will exceed those of self–matched fibers, since two different fibers will have variations in core size and eccentricity, even if the fibers are from the same manufacturer. If the two halves of the connector are from different manufacturers, the loss may increase. Connector losses will also increase if the fiber endfaces are not polished optically flat, since this will increase the reflection loss.

6.8: Optical Multiplexing

Splitters, combiners, and other optical multiplexers cause additional link loss. The link power transmission through an optical multiplexer can be divided into two effects: power division, and excess loss. Excess loss is the power loss due to light being entirely scattered out of the link.

In a splitter, power from an emitting fiber is divided among several receiving fibers. The amount of power directed into any one fiber is 1/N of the emitted power. An ideal two–way splitter, for example, puts 3 dB (1/2) of the input power into each of the two receiving fibers. Splitters are used in star networks and in bypass switches. When there are many output fibers, a passive splitter is replaced with an active splitter (electronic repeater or fiber amplifier) to ensure sufficient power is coupled into each receiving fiber.

In a splitter, optical power is lost at the joining due to changes in the number of modes as the physical size of the fiber changes in the splitter. Power is also lost due to reflections back into the emitting fiber. The power that is lost is not available to be divided among the receiving fibers, and is referred to as excess loss. Splitters can be made by twisting all of the receiving fibers together and fusing them together with the emitting fiber, as shown in Figure 6-12. Splitter design and manufacturing techniques are designed to minimize excess loss and equalize the power transfer into all of the receiving fibers.

A splitter data sheet normally specifies the excess loss, the maximum power ratio coupled from the emitting fiber into any one receiving fiber, and the power ratio reflected back into the emitting fiber.

Combiners take optical signals from several emitting fibers and combine them onto one receiving fiber. Combiners have excess loss due to losses at the physical joining and due to reflections. The data sheet normally specifies the excess loss, the maximum power ratio coupled from any one emitting fiber into the receiving fiber, and the power reflected back into another "emitting" fiber.

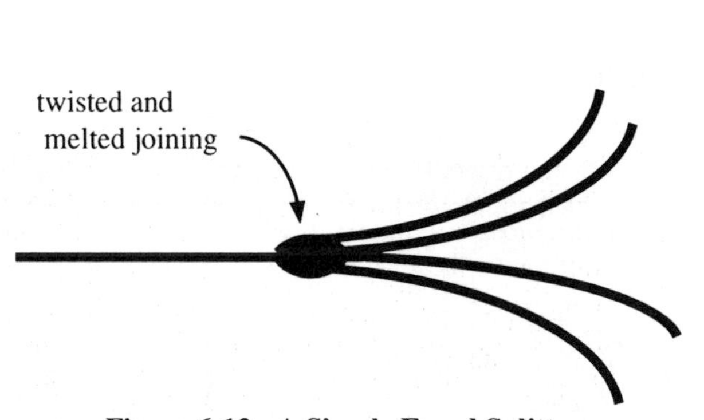

Figure 6-12: A Simple Fused Splitter.

EXAMPLE:

A 4–way splitter has an excess loss of 1.5 dB. Input power of 0.02 mW flows into the splitter. How much power leaves in each receiving fiber?

The input power is split 4 ways, so there is (0.02 / 4) mW or –17 dBm per receiving fiber. This power is further reduced by 1.5 dB due to excess loss. This results in each receiving fiber having –18.5 dBm of power.

Optical taps act as directional couplers. Taps are used to join stations to busses or to tap power for network monitoring. Bypass switches are used to route the optical signal around a station that is not actively in the network (for example, because its power supply has failed). A station is joined to a network through two taps: one to couple input optical signals from the network fiber, and one to couple outgoing optical signals. Figure 6-13 shows a bypass switch which uses two taps, to couple power from the main fiber into a bypass fiber and to couple the power back into the main fiber. Unlike splitters, taps can split power unevenly. Available coupling ratios range from 3 dB (50%) to 10 dB (10%). For example, a 10% tap coupler will leave 90% of the incoming optical power in the main fiber.

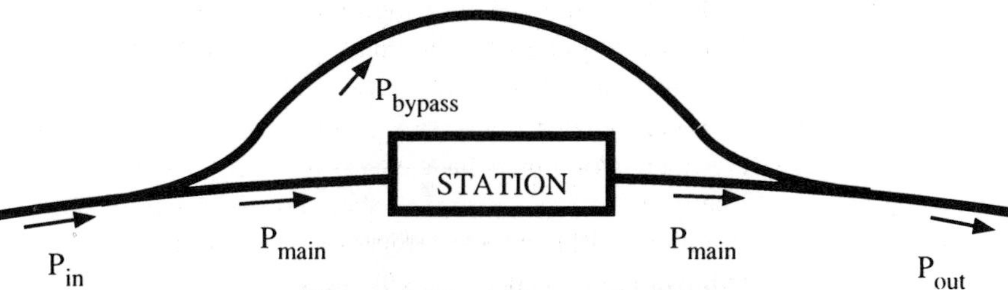

Figure 6-13: A Bypass Switch.

6.9: Fiber Amplifiers

In addition to losses, certain types of fibers can provide optical gain. Erbium–doped fiber amplifiers (EDFAs) reached commercial markets in late 1990, and are expected to allow almost unlimited repeater spacing for digital telecommunications systems.

Erbium–doped fiber amplifiers are typically designed to couple to single–mode fibers, and relatively short (less than 30 cm). Gains can be as high as 20 or 30 dB, depending on the optical signal's wavelength and the fiber amplifier length. Figures 6-14 and 6-15 show the gain of a commercial EDFA as a function of wavelength and input optical power, respectively.

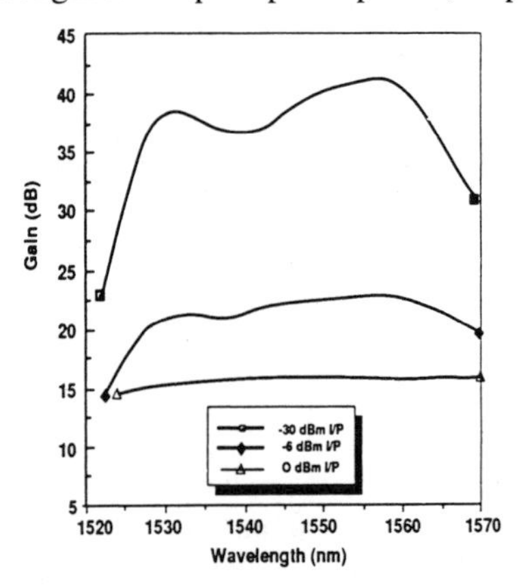

Figure 6-14: EDFA Gain vs. Wavelength.

Courtesy of BT&D Technologies.

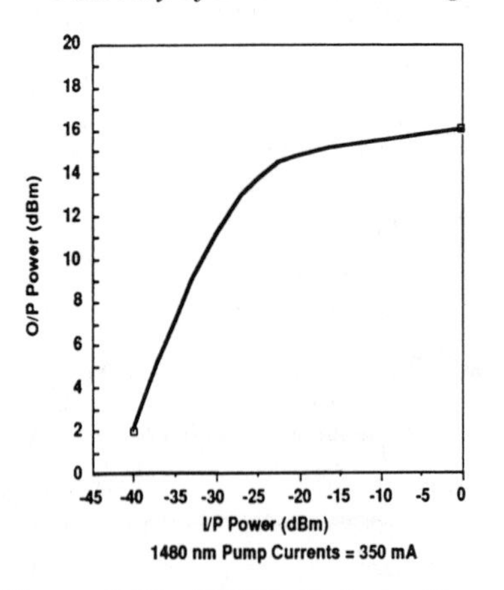

Figure 6-15: EDFA Gain vs. Power.

Courtesy of BT&D Technologies.

The fiber amplifier acts in a similar fashion to a laser, where stimulated emission results in optical gain. Unlike a laser, the optical power is not repeatedly reflected at the ends of the amplifier, but is coupled into the next section of fiber.

Power is provided to the fiber amplifier through an optical pump (similar to providing DC power to an op amp). The pumping wavelength is not the same as the optical signal wavelength, but is selected to efficiently deliver power to the amplifier. Because power is provided through an optical input, there are no electrical connections needed for the fiber amplifier. Electrical connections are required for the pump, which is usually a specialty lamp. Figure 6-16 shows a block diagram of an EDFA together with its support electronics and thermal control.

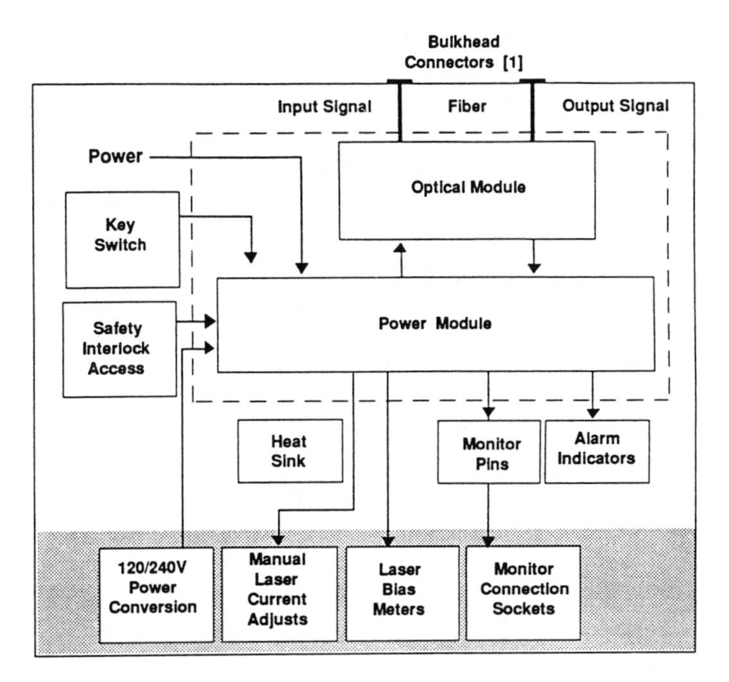

Figure 6-16: EDFA System Diagram.

Courtesy of BT&D Technologies.

Fiber amplifiers are used to provide gain to a link, on the order of 10–30 dB. Fiber amplifiers are used together with splitters; the splitting ratio is compensated by the gain of a fiber amplifier placed just before the splitter.

6.10: Summary

There are three basic fiber profiles: step–index, graded–index, and single–mode. Each fiber has advantages and disadvantages. Step–index fiber is easy to couple to, but has low bandwidth. Graded–index fiber is easy to couple to, has a medium bandwidth, and is the most expensive. Single–mode fiber, with less than 10–μm core diameter, is very difficult to couple to, but has high bandwidth and moderate cost.

Attenuation is the loss of optical signal strength down a length of fiber. Attenuation is highest in the 850–nm window. For long–haul communications, fiber is selected that has its lowest attenuation in the 1550–nm window. Bending losses will increase the observed attenuation. Sharp bends will snap the fiber.

The optical coupling loss decreases as the NA of the fiber increases. Optical coupling is on the order of 10–16 dB for LEDs and 3–5 dB for laser diodes. The coupling loss is further increased by source–fiber misalignment, particularly for single–mode fibers.

Other sources of loss include fiber endface reflections, which can be reduced with index matching materials, and misalignment between fibers at a connection. Losses are also caused by mismatches in fiber characteristics, such as core diameter, index profile, NA, core eccentricity, non–concentricity.

Connectors are used to provide connections that can be easily reconfigured, while splices provide lower loss permanent connections. Splitters, combiners, and other optical multiplexers also introduce loss in fiber power transmission.

Passive components always provide a power transfer ratio that is less than or equal to unity. Gain can be achieved with the use of erbium–doped fiber amplifiers, where DC power is provided to the amplifier through an optical pump.

6.11: Exercises

1. A fiber has a core index of 1.446 and a cladding index of 1.434. Find the index difference Δ and the NA.
2. A fiber has a core index of 1.446 and a critical angle of 12.8^{0}. Find the cladding index, the index difference Δ, and the NA.
3. A glass fiber has a core index of 1.456, a cladding index of 1.448, and a core diameter of 62.5 μm. Determine the number of modes, and the fraction of power propagating in the cladding when λ=0.85 μm.
4. A glass fiber has a V–parameter of 23.9. Determine the number of modes, and the fraction of power propagating in the cladding.

5. A glass fiber has a core index of 1.456, a cladding index of 1.448, and is a single–mode fiber when the optical carrier wavelength is 1300 nm.
   (a) What is the maximum core diameter?
   (b) Will this be a single–mode fiber if the wavelength is 820 nm?
   (c) Will this be a single–mode fiber if the wavelength is 1550 nm?
6. A fiber is to be selected for single–mode operation in the 850 nm window (800–900 nm). What is the maximum core diameter for such a fiber if the NA is 0.18?
7. A fiber having has a core diameter of 8 μm is to be used at 1300 nm. What is the maximum NA for this fiber to operate with a single mode ?
8. A step–index fiber has an NA of 0.20. What percentage of the power from a surface LED is coupled into the fiber?
9. Express the coupling loss in dB between an LED and a fiber having a numerical aperture of 0.20.
10. A graded–index fiber has an NA of 0.18. What percentage of the power from a surface LED is coupled into the fiber?
11. A step–index fiber couples 0.018 mW of power from an LED whose output power is 0.37 mW.
   (a) What is the NA?
   (b) What is the cladding index if the core index is 1.446?
12. A 50/125 step–index fiber has an NA of 0.23. What is the coupled power from a 75–μm diameter LED whose output power is 0.3 mW? Express this power in both mW and dBm.
13. An LED couples 0.23 mW into a step–index fiber of NA 0.18. How much power will be coupled into a step–index fiber of NA 0.20? Express this power in both mW and dBm.
14. A fiber has a coupled power of 0.25 mW, an attenuation of 10 dB/km, and a length of 0.4 km. Find the fiber output power. Express this power in both mW and dBm.
15. A fiber has a coupled power of 0.05 mW, an attenuation of 1 dB/km, and a length of 4 km. Find the fiber output power. Express this power in both mW and dBm.
16. An LED has an output power of –4 dBm. It is coupled to a 1 km piece of fiber having an NA of 0.21 and attenuation of 3 dB/km. What is the fiber output power?
17. A fiber has a required output power of 750 nW, an NA of 0.21, an attenuation of 10 dB/km, and a length of 0.7 km. What is the minimum acceptable source power, assuming an LED?

18. A single–mode fiber has an effective diameter of 8 μm and an NA of 0.12. Calculate the coupled power from (a) an LED whose diameter is 50 μm and (b) a laser of dimensions 0.2 x 4 μm, if each puts out 1 mW into air.
19. A 50/125 step–index fiber has an NA of 0.23 and a core index of 1.450. It carries a 1550–nm optical signal. A 2–cm bend is introduced into the fiber. Calculate the power transmission ratio, and express the power loss in dB.
20. A 50/125 graded–index fiber has an NA of 0.23 and a core index of 1.450. A 2–cm bend is introduced into the fiber. It carries a 1550–nm optical signal. Calculate the power transmission ratio, and express the power loss in dB.
21. Plot the power transmission ratio vs. bend radius for the fibers of Problem 6–19 and 6–20 for bend radii from 0 to 10 cm. Comment on (a) where the formula appears to give incorrect answers, (b) which type of fiber shows lower bending losses, and (c) at what radius bending losses become negligible for each fiber (less than 5% loss).
22. Plot the axial offset and separation losses (in dB) on a graph from 0 to 5 μm. Assume the connection is between two identical 50/125 step–index fibers having a core index of 1.445 and an NA of 0.23.
23. A connector holds two fibers such that the endface separation is 4 μm, the maximum axial offset is 1 μm, and the maximum angular offset is 2°. Calculate the loss in the connector if it is used to connect two 50/125 step-index fibers having a core index of 1.445 and an NA of 0.23.
24. A 62.5/125 step–index fiber having an NA of 0.23 is butt–connected without an air gap to a 50/125 graded–index fiber having an NA of 0.18. For the step–index as the emitting fiber, calculate (a) the NA mismatch transmission ratio, (b) the grading profile mismatch power ratio, (c) the size mismatch power ratio, (d) the total power transfer ratio.
25. A 62.5/125 graded–index fiber having an NA of 0.23 is butt–connected without an air gap to a 50/125 step–index fiber having an NA of 0.18. For the graded–index as the emitting fiber, calculate (a) the NA mismatch transmission ratio, (b) the grading profile mismatch power ratio, (c) the size mismatch power ratio, and (d) the total power transfer ratio.
26. Each tap on the bypass switch shown in Figure 6-13 has a 1–dB excess loss, and the bypass tap couples 10% of the incident optical power into the bypass fiber. There is negligible loss in the short bypass fiber. What is the output power if the station is down, when 0.2 mW enters?
27. An erbium–doped fiber amplifier has 15 dB of gain. If the incident power is 150 μW, what is the output power?

Chapter 7 Link Bandwidth

7.1: Link Transfer Function

The transfer function of a linear link is the product of the transfer functions of the individual (cascaded) components. A fiber optic link consists of a transmitter, a fiber with associated connectors and splices, and a receiver. The link, seen from the electrical interface, forms a current transformer. The resulting transfer function is:

$$I_{out} / I_{in} = (P_s / I_{in})(P_c / P_s)(P_o / P_c)(I_p / P_o)(I_{out} / I_p)$$

If the receiver includes a transimpedance gain stage, the transfer function becomes:

$$V_{out} / I_{in} = P_s / I_{in})(P_c / P_s)(P_o / P_c)(I_p / P_o)(V_{out} / I_p)$$

The general form of the transfer function is:

$$H(f) = \frac{A}{\sqrt{1 + \frac{f}{f_p}}}$$

where A is the DC system transfer ratio and f_p is the pole (3 dB) frequency. The previous chapters concentrated on the low frequency response; this chapter discusses the frequency response of fiber optic links.

As an optical signal travels along a fiber, the signal amplitude decreases and the pulse width increases, as shown in Figure 7-1. For each bit, there is an associated time bin which contains the signal energy for that bit. In a short link, the optical pulse has sharp edges, and logical "1" and "0" levels are clearly distinguishable. For intermediate lengths, the optical energy in the "1" bit bin spreads into the adjacent bin, raising the energy in the adjacent bin and decreasing the energy in the original bin. For long links, this spreading of energy across bin boundaries can reduce the energy difference between a logical "1" and "0" to the point that they can no longer be distinguished.

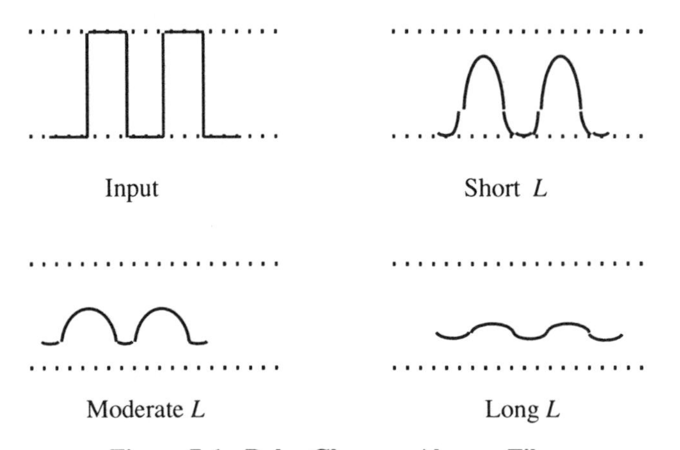

Figure 7-1: Pulse Changes Along a Fiber.

The pulse spreading or signal risetime in a fiber optic link can be modeled as a single–pole transfer function, with the pole frequency depending on the length of the fiber, the fiber profile, and the optical source spectral width. The signal–to–noise ratio at the end of the link must be sufficient to distinguish between "0" and "1" levels. The reduction in signal differential between the "1" and "0" levels for long links causes a decrease in SNR and an increase in the bit error rate. The link transfer function, which includes both the gain (attenuation) and pulse spreading (dispersion), is used in the same way as transfer functions for electrical networks.

The risetime increases with increasing fiber length, as shown in Figure 7-2. As with power transmission, a short length will have a stronger dependence on length due to the presence of higher order (leaky) modes. Beyond the equilibrium length, the fiber's contribution to link risetime is proportional to link length. When the mode distribution (between leaky and bound modes) reaches an equilibrium, as in long–haul links, pulse spreading in the fiber increases approximately proportional to $L^{1/2}$.

The notation for risetime will be the symbol t, and risetime per unit length will be τ. Other references use other notations, and care should be taken with equations. A dimensional or unit check will usually make it obvious which is meant in any given equation.

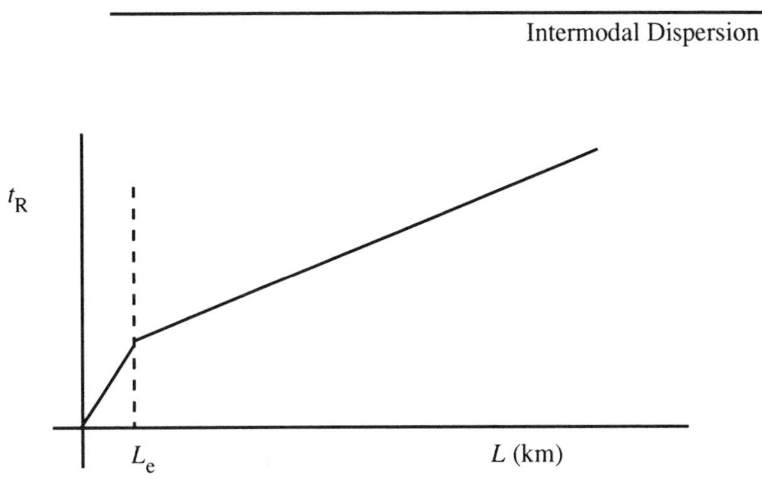

Figure 7-2: Risetime vs. Distance in a Fiber.

7.2: Intermodal Dispersion

Pulse spreading in the fiber, more commonly referred to as dispersion, limits the information bandwidth. Fiber dispersion has three major causes: modal dispersion, material or chromatic dispersion, and waveguide dispersion.

Modal dispersion is due to different rays (modes) having different arrival times at the end of the fiber. For a step–index fiber, the ray that travels in a straight line takes time L/v to arrive at the end of the fiber. The ray that travels at the critical angle (θ_c) takes a time $L/(\mathrm{v} \sin \theta_c)$ to arrive. The modal pulse spreading width is the difference of these two times:

$$tw_{\mathrm{mod}} = L / [\, \mathrm{v}\,(1 - 1/\sin \theta_c)\,] = n_1 \, \Delta \, L / c = (\, \mathrm{NA}^2 \,)\,(\, L \,) / (\, 2\, n_1\, c\,)$$

The risetime increases as NA increases for both step– and graded–index fibers. The modal dispersion is a factor of 10 to 100 lower in graded–index fiber than in step–index fiber, and is sensitive to the index profile grading (α). For both step– and graded–index fibers, there is a tradeoff with NA between risetime and coupling efficiency, since both increase with increasing NA. Each manufacturer will optimize the NA for a particular tradeoff of bandwidth and coupling efficiency.

The modal dispersion of a fiber is commonly given in terms of the distance–bandwidth (B*L) product. For a given fiber, the B*L product is independent of wavelength for most purposes. Fibers shorter than the equilibrium length may exhibit significantly lower B*L, due to higher order modes still present in the fiber. The bandwidth symbol B_0 is used for the bandwidth of a 1–km length of fiber, and has the same numerical value as B*L.

EXAMPLE:

A step–index fiber has a core index of 1.445, numerical aperture of 0.27, and a length of 2 km. What is the modal pulse spreading in the fiber?

$$tw_{mod} = (NA^2)(L)/(2n_1c)$$
$$tw_{mod} = [(0.27^2)(2\text{ km})] / [(2)(1.445)(3 \times 10^8\text{ m/sec})(1\text{ km} / 10^3\text{m})]$$
$$tw_{mod} = 168 \times 10^{-9}\text{ sec} = 168\text{ nsec}$$

EXAMPLE:

Calculate the bandwidth–distance product for the step–index fiber of the previous example.

t_{mod} = 168 nsec and $B = 2.08$ MHz when $L = 2$ km
B∗L = (2.08 MHz) (2 km) = 4.16 MHz–km

Typical B*L values for step–index fiber are 5–20 MHz–km. Graded-index fiber has higher B*L, typically 200–800 MHz–km. Single–mode fiber, carrying only one mode, does not suffer from modal dispersion.[1]

The modal dispersion tw_{mod} describes the optical pulse spreading due to modal effects. Optical pulse width can be converted to electrical risetime through the relationship:

$$t_{r,\,mod} = 0.44\ tw_{mod}$$

The ideal B*L predicts a linear increase in risetime with length in the absence of mode mixing. In practice, mode mixing decreases the length dependence. A number of empirical formulas have been developed that incorporate the effects of microbends and mode mixing, as well as other physical effects. One empirical formula used for link design is:

$$t_{r,\,mod} = 0.44\,L^q\ /\ B_o$$

where $t_{r,\,mod}$ is the signal risetime, the factor 0.44 is the conversion from optical pulse width to signal risetime,[2] and q describes the mode mixing effects. The factor q assumes that the length L is explicitly in kilometers. Local area network links have $0.7 < q < 1.0$, while long–haul links having an

[1] The B_o product specified on data sheets for single–mode fiber is waveguide dispersion.

[2] Assuming a Gaussian pulse shape.

equilibrium mode distribution have $q = 0.5$. Conservative designs use $q = 1$, while experienced designers use $q = 0.7$ when appropriate.[1]

EXAMPLE:

A step–index fiber has a core index of 1.445, a numerical aperture of 0.27, and a length of 2 km. What is the bandwidth distance product of the fiber? Assume $q = 1$.

$$t_{r,\,mod} = 0.44\, tw_{mod} = (0.44)\,(168 \text{ nsec}) = 73.9 \text{ nsec.}$$

$$B_o = 0.44\, L^q \,/\, t_{r,\,mod} = L^q \,/\, tw_{mod} = (2 \text{ km})^{1.0} / 168 \times 10^{-9}$$

$$B_o = 12 \text{ MHz–km.}$$

7.3: Intramodal Dispersion

Material and waveguide dispersion are due to the fiber material and the fiber shape, respectively. Material (chromatic) dispersion is caused by the dependence of index of refraction on wavelength, causing wavelengths λ and $\lambda + d\lambda$ to travel at slightly different speeds. Waveguide dispersion is caused by the difference in index of refraction between the core and cladding, resulting in a “drag” effect between the core and cladding portions of the power.

Material dispersion is commonly characterized by the dispersion coefficient D. The material pulse risetime is given by:

$$t_{mat} = -D\, \sigma_\lambda\, L$$

where D is the dispersion coefficient, σ_λ is the source spectral width, and L is the link length. Typical units for D are nsec/nm/km or psec/nm/km. The dispersion per unit distance is:

$$\tau_{mat} = -D\, \sigma_\lambda$$

For pure silica glass, the dispersion coefficient can be fit by the equation

$$D = 1320\, e^{-2.805\,\lambda} - 40 \qquad \text{D in psec/nm/km} \quad \lambda \text{ in } \mu\text{m}$$

---

[1] The variation of q with length will depend on the fiber grading profile, and on the number of macrobends and microbends in the fiber. The actual value of q is not easily predicted from first principles.

This equation is a reasonable fit over the range 0.75–1.7 μm. The actual dispersion coefficient depends on the glass doping, and will vary accordingly. Fiber data sheets may give a value of D at one wavelength, a plot of D vs. λ, or a maximum value over a wavelength window.

The dispersion coefficient is strongly dependent on wavelength. For pure silica glass fibers, the minimum in $|D|$ occurs near 1250 nm. This means that the lowest material dispersion occurs in the 1300–nm window, while the best attenuation occurs in the 1550–nm window. In the 1300–nm window, the dispersion will be sensitive to the source's center wavelength, since a small change in wavelength produces a large percentage change in the dispersion coefficient.

Two special grades of single–mode fiber have been developed by fiber manufacturers to address long links operating in these two windows. The first type is "dispersion–flattened" fiber, in which the slope of the D vs. λ curve has been flattened near the zero crossing. The second type is "dispersion–shifted" fiber, where the zero crossing has been moved to the 1550–nm window. The highest performance systems use a fiber that has been both dispersion flattened and dispersion shifted, so that the maximum value of D throughout the 1550–nm window is as small as possible.

EXAMPLE:

A step–index fiber has a core index of 1.445, a numerical aperture of 0.27, and a length of 1 km. Calculate the material dispersion, and compare it to the modal dispersion. Assume an LED source having λ = 800 nm and $\sigma_\lambda = 50$ nm.

(a) At 800 nm, D = 0.1 nsec/nm/km.

$\tau_{mat} = -D\,\sigma_\lambda = -(0.1 \text{ nsec / nm / km})\ (50 \text{ nm}) = -5 \text{ nsec / km}$

(b) For $q = 1$, $\tau_{mod} = 84 \text{ nsec / km}$

Here, the magnitude of τ_{mod} is 16 times greater than that of τ_{mat}.

EXAMPLE:

Calculate the material dispersion of the step–index fiber of the previous example, and compare it to the modal dispersion. Assume a laser diode source having λ = 800 nm and $\sigma_\lambda = 2$ nm.

(a) At 800 nm, D = 0.1 nsec / nm / km.

$\tau_{mat} = -D\,\sigma_\lambda = -(0.1 \text{ nsec / nm / km})\ (2 \text{ nm}) = -0.2 \text{ nsec / km}$

(b) As before, $\tau_{mod} = 84 \text{ nsec / km}$

Here, the magnitude of τ_{mod} is 420 times greater than that of τ_{mat}.

EXAMPLE:

Calculate the material dispersion of the step–index fiber of the previous example, and compare it to the modal dispersion. Assume an LED source having $\lambda = 1550$ nm and $\sigma_\lambda = 50$ nm.

(a) At 1550 nm, D = – 0.3 nsec/nm/km.
$\tau_{mat} = -D\ \sigma_\lambda = -(-0.3 \text{ nsec / nm / km})\ (50 \text{ nm}) = +15 \text{ nsec / km}$
(b) As before, $\tau_{mod} = 84$ nsec / km
Here, τ_{mod} is 5.6 times greater than τ_{mat}.

Material dispersion is large at short wavelengths, where D is large. Material dispersion is larger for LED than for laser sources, since LEDs have larger spectral widths. Material dispersion can be decreased by moving to the zero–dispersion window, and by replacing LEDs with lasers.

Waveguide dispersion is significant only in fibers carrying fewer than 5–10 modes. Since multimode optical fibers carry hundreds of modes, they will not have observable waveguide dispersion. Single–mode fibers, on the other hand, exhibit waveguide dispersion that is comparable to the material dispersion.

The waveguide dispersion depends on the core diameter, the NA, the V-number, and the optical wavelength. The pulse risetime due to waveguide dispersion can be estimated from:

$$t_{WG} = -\left(\frac{n_2 L \Delta \sigma_\lambda}{c\lambda}\right)(z)$$

where z is a parameter that depends on the V–number of the single–mode fiber.[1] The parameter z has a maximum value of 1.0 near V = 1.3. When the fiber is operated near the cutoff wavelength, $z = 0.2$ is usually a good approximation. For V > 3, z rapidly approaches zero. Graphs of z can be found in texts on fiber optics and optical waveguide theory.

[1] The parameter z comes from the Bessel equation .solution to the wave equations, and is defined as:

$$z = V \frac{\partial^2 (V b)}{\partial V^2}$$

EXAMPLE:

A single–mode fiber is operated with V = 2. The core index is 1.450 and the cladding index is 1.440. Estimate the waveguide dispersion per km for an optical carrier having a wavelength of 1320 nm and a spectral width of 2 nm.

Since V is near 2, use the approximation $z = 0.2$.

$\tau_{WG} =$
$-[(1.440)(10^3 \text{ m/km})(0.0069)(2 \text{ nm})(0.2)] / [(3 \times 10^8 \text{ m/sec})(1320 \text{ nm})]$
$\tau_{WG} = -10 \times 10^{-12} \text{ sec / km} = -0.010 \text{ nsec / km}$

Comparing the results of this example with the results of previous examples shows that the waveguide dispersion will be much smaller than material dispersion if the optical source is an LED. Waveguide dispersion will be smaller than material dispersion at short wavelengths as well.

Both material and waveguide dispersion contribute to intramodal dispersion, since both occur even when there is only one mode in the fiber. The total intramodal dispersion is the direct sum of the material and waveguide dispersions:

$$\tau_{intra} = \tau_{mat} + \tau_{WG}$$

The material dispersion coefficient D changes sign, and there is a wavelength at which the intramodal dispersion crosses zero. This is not the same wavelength as where D crosses zero, since the waveguide dispersion is never zero for a single–mode fiber. A long–haul link will have its best performance at the wavelength at which the intramodal dispersion is zero.

In actual practice, the intramodal dispersion of a link is never exactly zero. This is because the optical source has a finite (non–zero) spectral width, so that some portion of the optical power may experience a D sufficiently different from the optimal one as to contribute a small amount of intramodal dispersion. Furthermore, the actual center wavelength of a laser diode cannot be controlled sufficiently to keep it at exactly the zero intramodal dispersion wavelength over all environmental variations.[1]

7.4: Total Fiber Dispersion

The modal dispersion in the fiber is also referred to as the intermodal dispersion, meaning that it is due to an effect between different modes. The

[1] In the field, as opposed to in the lab.

total dispersion in the fiber is the RMS sum of the (uncorrelated) intermodal and intramodal dispersions:

$$\tau_{rf} = (\tau^2_{intra} + \tau^2_{inter})^{1/2} \quad \text{(per unit length)}$$

$$t_{rf} = (t^2_{intra} + t^2_{inter})^{1/2} = (t^2_{mod} + [t_{mat} + t_{WG}]^2)^{1/2}$$

For a multimode fiber, the waveguide dispersion is zero, and

$$t_{rf, MM} = [(D\,\sigma_\lambda\, L)^2 + (0.44\; L^q / B_o)^2]^{1/2}$$

For single–mode fiber, the intermodal dispersion is zero, and

$$t_{rf, SM} = \left| -D\,\sigma_\lambda\, L \;-\; \left(\frac{n_2 L \Delta \sigma_\lambda}{c\lambda}\right)(z)\right|$$

The fiber dispersion may be calculated on either a total basis, or on a per–unit–length basis. The difference is obvious if unit analysis is made. Extreme care must be taken in link design to ensure that the total dispersion over the entire length of the fiber is used. The total dispersion in the fiber is determined from the per–unit–length value (assuming $q = 1$):

$$t_{rf} = \tau_f\, L$$

Typically, the units of t_{rf} are nsec and the units of τ_f are nsec/km. Care must be taken include appropriate mode mixing in the intermodal term.

EXAMPLE:

A fiber has 7 nsec/km of intermodal dispersion, 2 nsec/km of material dispersion, and no waveguide dispersion. Find the total dispersion in a 5 km span.

(a) $\tau_f = (7^2 + 2^2)^{1/2}$ nsec / km $=$ 7.3 nsec / km

(b) Total $t_{rf} = \tau_f\; L =$ (7.3 nsec/km) (5 km) $=$ 36.4 nsec.

7.5: Link Bandwidth

The link bandwidth depends on the transmitter and receiver risetimes as well as the fiber dispersion. As discussed in Chapter 2, the risetime is related to the (analog) bandwidth:

$$B = 0.35 / \tau$$

The link risetime is the RMS sum of the component risetimes:

$$t_{link} = \sqrt{t_{tx}^2 + t_{rf}^2 + t_{rx}^2}$$

EXAMPLE:

A 2–km fiber having B_o = 440 MHz–km is connected between a transmitter having a 2 nsec risetime and a receiver having a bandwidth of 100 MHz. What is the link risetime, assuming negligible material dispersion?

t_{rf} = (0.440) (2 km) / 440 MHz–= 2.0 nsec
t_{rx} = 0.35 / 100 MHz = 3.5 nsec

$$t_{link} = \sqrt{2^2 + 2^2 + 3.5^2} = 4.5 \text{ nsec}$$

EXAMPLE:

A 2–km fiber having B_o = 44 MHz–km is connected between a transmitter having a 13 nsec risetime. If the link bandwidth is to be at least 10 Mbit/sec, what is the minimum acceptable receiver risetime, assuming negligible material dispersion?

t_f = (0.440) (2 km) / 44 MHz–km = 20 nsec
t_{link} = 0.7 / 10Mb/sec = 70 nsec

$$t_{rx} = \sqrt{t_{link}^2 - [t_{tx}^2 + t_f^2]} = \sqrt{70^2 - [13^2 + 20^2]}$$

t_{rx} = 65.8 nsec

It is interesting to note that the longest risetime term will generally dominate the link risetime. In the above example, even though the transmitter and fiber risetimes are each about 20% of the link risetime, the allowed receiver risetime was over 90% of the link risetime. This is a direct result of the fact that risetimes add in an RMS manner.

7.6: Non–linear and Harmonic Effects

A fiber optic link is generally not strictly linear. For example, a laser diode shows a strong non–linear threshold effect in the P–I characteristics. Smaller non–linearities are seen in both LED and laser transmitters, even when operated above threshold. Non–linearities may also occur in the receiver circuits.

The non–linearities in the link transfer function produce a number of observable effects in the link performance. The linearity of an analog system depends directly on the linearity of the link. Analog system linearity can be improved by pre–distortion of the signal at the transmitter or equalization at the receiver to compensate for non–linearities, making the overall system transfer function linear. Pre–distortion and equalization are used primarily in high–performance video systems.

Non–linear transfer functions result in harmonic generation, with output signal strength appearing at multiples (harmonics) of the original frequency. Thus a 10 MHz input signal will produce an output with most of the signal power at 10 MHz, but with some of the power appearing at 20 MHz, 30 MHz, etc. The amount of power in each of the harmonics will depend on the non-linear response of the link.

Non–linear functions also cause intermodulation product distortion. When an input signal contains information at two or more frequencies, the output will contain some signal power at sums and differences of the input frequencies. Thus, an input with signals at ω_1 and ω_2 will have output power at the following frequencies:

$$\omega_1, 2\omega_1, 3\omega_1, \ldots, n\omega_1$$

$$\omega_2, 2\omega_2, 3\omega_2, \ldots, m\omega_2$$

$$|\omega_1 - \omega_2|, |\omega_1 + \omega_2|, |\omega_1 - 2\omega_2|, |\omega_1 + 2\omega_2|, |2\omega_1 - \omega_2|, |2\omega_1 + \omega_2|, \ldots, |n\omega_1 - m\omega_2|$$

The designer is concerned only with those harmonic and intermodulation terms which are inside the frequency bandwidth of the signal (the link bandwidth). With AC receiver coupling, a link appears as a bandpass filter in its frequency response. The typical link has both a lower and upper cutoff frequency. Many links are designed to have a 1–octave bandwidth, where the highest frequency is twice the lowest frequency. For many analog links, the first 2 or 3 cross terms and the first harmonic are inside the link bandwidth.

Another source of distortion is signal dependent noise. Signal dependent noise is caused by an echo of the input arriving during at a later time. Because the noise is identical with a valid signal, it lies in the same frequency band as the original signal. Signal dependent noise is a result of reflections within the link. For a short link, the reflections from the receiver endface of the fiber, back to the input end of the fiber, and finally back to the output end, may arrive with enough signal strength to appear as a significant noise factor. The reflected power will depend on both the reflection at the endfaces and the fiber attenuation. This effect is not significant for longer links, since the reflected signal travels 3 *L* while the original signal travels *L*. However, this effect may be significant if there are short jumpers.[1]

Signal dependent noise can be large in laser transmission systems, where it is also referred to RIN or reflection–induced noise. Because the laser acts as a light amplifier, it will amplify any light reflected back into the laser from the fiber. This is a significant effect in video systems and high data rate systems having jumper cables. Reflections back into a laser can be prevented by use of anti–reflection coatings at all connectors at the transmitting end of the link, avoidance of jumper cables, use of optical isolators at the laser connector, or use of specially designed connectors which reduce back reflections.

7.7: System Tradeoffs

The performance of a fiber optic link depends on the fiber index profile, core diameter, and NA. The optical pulse spreading within the fiber also depends on the wavelength and spectral width of the optical source.

The highest performance is achieved with a 1550–nm laser coupled to a dispersion shifted single–mode fiber. The laser provides a small spectral width, while the single–mode fiber has no intermodal dispersion and a minimum intramodal dispersion near 1550 nm. Dispersion shifted fiber provides minimum attenuation in the same operating window as minimum attenuation for silica fibers.

The lowest performance is achieved with a 820–nm LED coupled to a step–index fiber. An LED has a large spectral width. Step index fiber has the highest modal dispersion, and the fiber has a large dispersion coefficient in the short–wavelength window.

Intermediate performance is achieved with a 1300–nm LED and graded-index fiber, or an 850–nm laser and single–mode fiber. At 1300 nm, the large spectral width of the LED and the smaller dispersion coefficient of the fiber provide comparable material dispersion to that of a laser operating in a region of higher dispersion coefficient.

When fibers of different grades are connected, each fiber section will have a different bandwidth. The lowest performance section will most limit the system bandwidth. A series connection of fiber links can be analyzed in the same fashion as a cascaded electronic network, as long as the transfer

---

[1] If T is the bit period, then a jumper with a length less than v / (10T) will not cause a problem, while one with a length of v/3T or longer is of concern.

function for each section is known. As with electronic networks, the system transfer function is given by the product of the individual section transfer functions.

7.8: Summary

The link risetime is the RMS sum of the individual component risetimes. The transmitter and receiver risetimes depend on both the optoelectronic components and the associated circuits. The fiber dispersion depends on the transmitter wavelength and spectral width.

There are three contributions to pulse spreading or dispersion in fibers. Modal dispersion is of concern for all multimode fibers. Modal dispersion is greater in step–index fibers than in graded–index fibers. Waveguide dispersion is of concern only for single–mode fibers (which may become multimode fibers if operated below the cutoff wavelength).

Material dispersion is determined by the properties of the optical signal. The material dispersion coefficient depends on the optical wavelength. The material dispersion is also directly proportional to the spectral width. Material dispersion can be minimized by operating near the zero–dispersion wavelength of 1250 nm for silica fiber and 1550 nm for dispersion shifted fiber, and by using laser sources. Material dispersion is of concern in both multimode and single–mode fibers.

The risetime of an optical signal propagating in a fiber is the RMS sum of the intermodal and intramodal risetimes. The intramodal risetime is the sum of the material and waveguide risetimes.

A fiber is often characterized by its distance–bandwidth product, B*L or B_0. This parameter characterizes either the modal or waveguide dispersion, depending on the type of fiber. Step–index fiber has typical values around 10 MHz–km; graded–index fiber has 200–800 MHz–km. Single–mode fiber, which does not have modal dispersion, has waveguide dispersion which limits the fiber to around 100 GHz–km.

In step–index fiber, the primary limitation is intermodal dispersion. In graded–index fiber, material dispersion is dominant at short wavelengths and intermodal at long wavelengths. In single–mode fiber, material dispersion dominates at short wavelengths, while material and waveguide dispersion have comparable magnitudes at longer wavelengths.

Non–linearities in the link transfer function lead to harmonic and intermodulation effects. These cause portions of the signal power to be frequency shifted. If the shifted frequency is in the link passband, then the harmonic and intermodulation products contribute to the link noise. Similarly, reflection–induced noise, caused by reflections in the link that arrive at the receiver with enough power to be amplified, contributes to link noise.

The best link performance is achieved by operating in the highest wavelength window with a laser coupled to a single–mode fiber. The lowest performance is achieved by operating in the shortest wavelength window with an LED coupled to a step–index fiber.

7.9: Exercises

1. A link has a DC gain $A = 3 \times 10^{-4}$ and a pole frequency of 100 MHz.
   (a) Plot the output amplitude vs. frequency using a linear scale for amplitude and a log scale for frequency.
   (b) Make a Bode plot of the frequency response.
2. (a) Calculate the modal dispersion for a 1.5 km step–index fiber having a core index of 1.448 and an NA of 0.21.
   (b) Can this fiber be used to transmit NRZ data at 10 Mbit/sec over this distance? Explain why or why not.
3. (a) Calculate the modal dispersion for a 1.5 km graded–index fiber having B*L of 200 MHz–km. (b) Can this fiber be used to transmit NRZ data at 10 Mbit/sec over this distance? Explain why or why not.
4. Plot the material dispersion D for pure silica from 0.7 to 1.7 μm.
5. What is the percent change in material dispersion if the center wavelength changes from 820 to 850 nm?
6. Calculate the material dispersion for a 1.5 km fiber operating at 820 nm if the source spectral width is 37 nm.
7. Estimate the waveguide dispersion in a single–mode fiber having a core index of 1.440 and $\Delta = 0.007$, if the fiber's cutoff wavelength is 1200 nm and the laser source has a wavelength of 1335 nm and a spectral width of 1.6 nm.
8. A fiber has a modal dispersion of 20 nsec/km, material dispersion of –16 nsec/km, and no waveguide dispersion. What is the total dispersion if the fiber is 2.4 km long?
9. A fiber has a modal dispersion of 5 nsec/km, material dispersion of –1.6 nsec/km, and waveguide dispersion of 2.7 nsec/km. What is the total dispersion if the fiber is 2.4 km long?
10. Compare (tabulate) the modal, material, waveguide dispersion, and total dispersion for 1 km links:
    (a) MM fiber: $B_o = 10$ MHz–km, LED: $\lambda = 850$ nm, $\sigma_\lambda = 45$ nm.
    (b) MM fiber: $B_o = 400$ MHz–km, LED: $\lambda = 850$ nm, $\sigma_\lambda = 45$ nm.
    (c) MM fiber: $B_o = 8$ MHz–km, LED: $\lambda = 850$ nm, $\sigma_\lambda = 45$ nm.

11. Compare (tabulate) the modal, material, waveguide dispersion and total dispersion for 1 km links:
    (a) MM fiber: $B_o = 10$ MHz–km, LED: $\lambda = 850$ nm, $\sigma_\lambda = 45$ nm.
    (b) MM fiber: $B_o = 400$ MHz–km, LED: $\lambda = 850$ nm, $\sigma_\lambda = 45$ nm.
    (c) SM fiber: $n_1 = 1.446$, NA = 0.18, LED: $\lambda = 850$ nm, $\sigma_\lambda = 45$ nm.
12. Compare (tabulate) the modal, material, waveguide dispersion and total dispersion for 1 km links:
    (a) MM fiber: $B_o = 100$ MHz–km, laser: $\lambda = 850$ nm, $\sigma_\lambda = 2.5$ nm.
    (b) MM fiber: $B_o = 100$ MHz–km, LED: $\lambda = 850$ nm, $\sigma_\lambda = 45$ nm.
13. Compare (tabulate) the modal, material, waveguide dispersion and total dispersion for 1 km links:
    (a) MM fiber: $B_o = 400$ MHz–km, LED: $\lambda = 1550$ nm, $\sigma_\lambda = 85$ nm.
    (b) MM fiber: $B_o = 400$ MHz–km, laser: $\lambda = 850$ nm, $\sigma_\lambda = 2.5$ nm.
14. Compare (tabulate) the modal, material, waveguide dispersion and total dispersion for 1 km links:
    (a) SM fiber: $n_1 = 1.446$, NA = 0.18; LED: $\lambda = 1550$ nm, $\sigma_\lambda = 85$ nm.
    (b) SM fiber: $n_1 = 1.446$, NA = 0.18; laser: $\lambda = 850$ nm, $\sigma_\lambda = 2.5$ nm.
15. A link has a transmitter risetime of 5 nsec, a receiver risetime of 2 nsec, and a fiber dispersion of 4 nsec. What is the link risetime?
16. A link has a transmitter risetime of 2 nsec, a receiver risetime of 1 nsec, and a fiber dispersion of 3 nsec. What is the link risetime?
17. A link is to transmit 10 Mb/sec Manchester encoded data. The available transmitter and receiver have risetimes of 2 nsec each. If the fiber has a dispersion of 50 nsec/km, what is the maximum length of the link?
18. A link is to transmit 125 Mb/sec NRZ data. The available transmitter and receiver have risetimes of 1.5 nsec each. If the fiber has a dispersion of 10 nsec/km, what is the maximum length of the link?
19. A link has fiber endface reflections of 60%. The fiber length is 0.2 km and it has an index of refraction of 1.445 and an attenuation of 3 dB/km.
    (a) What is the amplitude of the reflected power that arrives at the output compared to the original input power in dB?
    (b) What is the time delay between the arrival at the output of the original signal and the reflected signal? (Hint: draw a diagram showing all travel distances.)

Chapter 8 Link Analysis

Analysis predicts the performance of a set of components connected to form a link. The predicted performance is compared to the desired performance. If the performance is not acceptable, the analysis is used to identify the problem areas. If the performance is far greater than required, then lower cost components may be evaluated, seeking to lower the system cost.

The first step in link analysis is to define the system requirements. A simple check list is shown in Table 8-1. Link distance is determined by the route the fiber optic cable must traverse. The required system risetime is calculated from the data rate and the coding scheme, as discussed in Chapter 2. The data rate and BER (or SNR) is determined by system designers. The coding scheme may be specified either by system requirements (for example, 4B/5B for FDDI) or by the link designer.

Table 8-1: Link Checklist

Link distance :	________
Fiber length :	________
Data rate :	________
Coding scheme :	________
System t_r :	________
BER or SNR :	________
Design margin :	________
Additional design constraints :	________

The fiber length is frequently longer that the straight–line distance, since cables generally cannot be run in a straight line between the two endpoints. For example, indoor cable runs usually are placed in raceways in the ceiling or under the flooring, or in walls and shafts. The physical link distance is the distance measured through these cable runs, rather than across the center of each room.

As an example of calculating the fiber length, consider a link between two islands. The point–to–point distance is measured on a flat map. The cable must be laid underwater along the sea floor, where it is difficult to make accurate measurements. However, it is possible to estimate the physical cable length from the map–measured distance. If the designer does not properly estimate the physical cable length, the cable will not reach between the islands.

The design margin and other system constraints are determined through system reliability and data quality considerations. For example, a broadcast video signal has an SNR requirement set by the broadcast industry. The design margin and other design constraints are set by the required operating life and environmental conditions. When a long operating life is required, it is desirable to operate the transmitter at a lower output level, different from the nominal transmitter output level. A long operating life may also require an increased design margin to allow for anticipated repair splices over the life of the system.

Copies of the link checklist and power and risetime budget sheets are contained in Appendix H. These sheets list only the major contributions to the link performance. Additional losses, such as excess loss in combiners or splitter power division, must be incorporated if these components are used.

8.1: Components

Link analysis begins with consideration of the active modules. The key transmitter parameters are the optical output power and risetime. The key receiver parameters are the sensitivity and risetime. A module specification sheet will usually provide the typical risetime (or bandwidth) and minimum power (output or sensitivity). The transmitter power and receiver sensitivity influence the link power budget, while the transmitter and receiver risetimes influence the link risetime budget.

The attenuation of the optical fiber determines the signal loss along the link. The total attenuation along the length of the fiber influences the link power budget. The fiber bandwidth depends on the operating wavelength, the transmitter optical spectral width, and the fiber profile. The fiber bandwidth influences the link risetime budget.

Connectors cause signal loss due to physical misalignment between the two fiber sections. Additional loss is associated with connections between fibers of different characteristics. Splices also cause signal loss. The losses of the connectors and splices influence the link power budget. Connectors and splices do not directly affect the link risetime budget, although mode mixing at the fiber joining may influence fiber bandwidth.

Fiber amplifiers provide signal gain within the fiber link. Fiber amplifiers are the only source of gain in the link. The gain of a fiber amplifier will have the opposite sign of the loss terms in the link power budget.[1]

A design margin is incorporated in the design, and is treated as an additional loss. The design margin accounts for the variation of components and degradation with time. For example, if the nominal performance of a transmitter is used in the link power budget, then a design margin of 3 dB is typically allowed for the aging effects on the output power. Another 1 dB is allowed for receiver sensitivity degradation due to aging. Both the transmitter and receiver will show some variation of performance with temperature as well. Fiber shows little aging effect, unless there is an unanticipated humidity problem (anticipated ones are addressed through proper cable specifications). Connectors show variations in connection loss both with assembly variations and over connect / disconnect cycles; the amount of variation depends on the type of connector. Finally, the design margin must address additional splices and bending losses that will add to the loss during and after installation.

The link margin will depend on whether worst–or nominal component characteristics are used in the link power budget. A general rule of thumb is to allow 3–4 dB for the effects of component aging if nominal characteristics are used. Approximately 3–6 dB is used to cover manufacturing variations, environmental (temperature) sensitivity, microbending, and other unpredictable losses. These allowances total 3–10 dB of power margin.

The actual link margin allowed in a particular design will vary between designers. An experienced designer who has done a similar design will have information on how the actual installed link compared to the link analysis, and will leave more (or less) margin accordingly. When there is data on manufacturing variations (such as transmitter aging characteristics), then the worst-case component characteristics can be predicted with greater assurance, and the design margin adjusted accordingly.

8.2: Link Power Analysis

The link power budget is used to determine if there is sufficient power at the receiver to achieve the desired signal quality (SNR or BER). The received

---

[1] Remember that the convention is to use loss as positive dB, so that gain, being opposite, must be negative dB.

power is the transmitter optical output power minus all of the link losses and the design margin.

EXAMPLE:

A transmitter has an output power of 0.1 mW. It is used with a fiber having NA = 0.25, attenuation of 6 dB/km, and length 0.5 km. The link contains two connectors of 2–dB average loss. The receiver has a minimum acceptable power (sensitivity) of –35 dBm. The designer has allowed a 4–dB margin. Calculate the link power budget.

$P_s = 0.1$ mW $= -10$ dBm
Coupling (NA) loss $= -10 \log (\mathrm{NA}^2) = -10 \log (0.25^2) = 12$ dB
Fiber loss $= \alpha L = (6$ dB/km$)(0.5$ km$) = 3$ dB
Connector loss $= 2\,(2$ dB$) = 4$ dB
Design margin $= 4$ dB
$P_{out} = P_s - \Sigma\,(\text{losses}) = -10$ dBm $-$ [12 dB + 3 dB + 4 dB + 4 dB]
$P_{out} = -33$ dBm

Since $P_{out} > P_{min}$, this system will perform adequately over the system operating life.

EXAMPLE:

A transmitter has an output power of 0.1 mW. It is used with a fiber having NA = 0.25, attenuation of 6 dB/km, and length of 1.5 km. The link contains two connectors of 2–dB average loss. The receiver has a minimum acceptable power (sensitivity) of –35 dBm. The designer has allowed a 4–dB margin. Calculate the link power budget.

$P_s = 0.1$ mW $= -10$ dBm
Coupling (NA) loss $= -10 \log (\mathrm{NA}^2) = -10 \log (0.25^2) = 12$ dB
Fiber loss $= \alpha L = (6$ dB/km$)(1.5$ km$) = 9$ dB
Connector loss $= 2\,(2$ dB$) = 4$ dB
Design margin $= 4$ dB
$P_{out} = P_s - \Sigma\,(\text{losses}) = -10$ dBm $-$ [12 dB + 3 dB + 9 dB + 4 dB]
$P_{out} = -39$ dBm

Since $P_{out} < P_{min}$, this system will *not* perform adequately over the system operating life. In examining the loss budget, it is noted that the largest loss is associated with the fiber attenuation. Possible ways to improve the design include using a different wavelength (where the fiber has a lower attenuation) or using a source with a greater output power.

The specifications for components are usually given for one set of operating conditions. For example, receiver sensitivity may be specified at 10 Mbit/sec, BER = 10^{-9}, T = 25°C, and V_{cc} = 5 V. Receiver sensitivity is most commonly given for a BER of 10^{-9}. Some component specifications will be given over the entire operating range (for example, 0–70°C). In other cases, it is necessary to apply derating factors to correct for these factors. Component data sheets may include derating factors, or they can be determined through measurements if components are available for testing.

Derating factors are most commonly applied to receiver sensitivity for dependence on data rate and BER. Derating factors can also be used for dependence on wavelength, temperature, DC bias voltage, and aging. One of the most common instances for applying a derating factor is when the system BER is more stringent than for the nominal sensitivity on the specification sheet.

Figure 8-1 shows BER vs. received power for a digital receiver. The more stringent the BER requirement, the greater the required power at the receiver. A higher speed link will require more received power to achieve a given BER. Typically, the slope of the line varies only slightly over data rate and environmental conditions, so that a rule–of–thumb approximation of 2 dB/decade (receiver power vs. BER) is commonly used. Once a specific receiver has been selected, it may be desirable to verify the derating factor with actual measurements if the factor has a significant impact on link performance.

Rules–of–thumb also exist for derating received power vs. data rate. The required receiver power increases with higher data rate, and this rate of change depends on the receiver design. A rule–of–thumb value may be provided for a particular receiver design by the manufacturer, or it may be developed from measured results on actual receiver modules.

The required receiver power will also depend on the DC bias applied to the transmitter diode. The ratio of the transmitter peak output power to the minimum output power is called the extinction ratio (typically zero for an LED and nominal threshold for a laser). The rated sensitivity for a receiver is given for a particular extinction ratio. Changing the DC bias point of the transmitter diode will change the extinction ratio and the receiver sensitivity. Extinction ratio dependence is a function of the receiver design. This dependence can be calculated, or characterized against data sheets or measured characteristics for a particular receiver.

Received power requirements also depend on the data encoding, particularly since this may change the average power level in a similar fashion as extinction ratio. As seen in Figure 8-1, changing the coding scheme changes the BER for a given power, and power for a given BER.

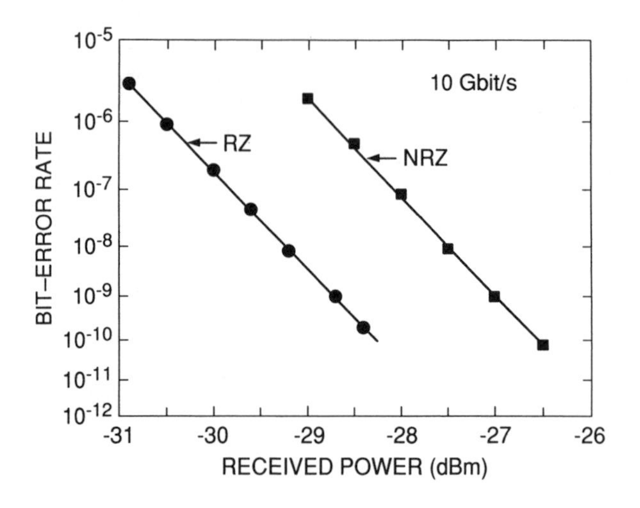

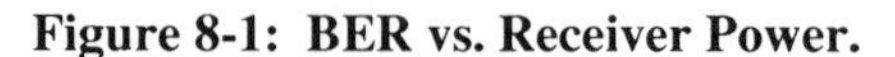

Figure 8-1: BER vs. Receiver Power.

Reprinted from A. Gnauck et. al. "A Transimpedance APD Optical Receiver Operating at 10 Gb/s", *IEEE Photonics Technology Letters*, May 1992. Copyright 1992 by IEEE. Reprinted with permission.

Receiver sensitivity is dependent on bias and temperature, and should be characterized over the entire operating range. Derating factors for bias and temperature are not generally used, since a transmitter is normally specified to operate in a link over a specified operating range.

These rules–of–thumb for derating factors are usually given in one of two forms. The change in received power requirements may be given vs. the change in operating parameters. Alternatively, the change in BER may be given rather than a change in power. The rule of thumb of 2 dB/decade of BER change allows one to convert between the two.

8.3: Digital Links

A digital link uses digital transmitters and receivers, selected to have the desired electrical interface (e. g. TTL levels). The transmitter and receiver are generally selected as a set to ensure compatible encoding and decoding schemes. The TX/RX set will generally have both upper and lower data rate cutoffs, since AC coupling is used in the receiver to improve the BER.

The TX/RX set specifications commonly give the power loss budget between the transmitter and receiver, as well as a specification on the grade of fiber for which the components are designed. For example, a TX/RX set may specify a 10–dB power budget over 62.5/125 step–index fiber having an NA of 0.2. Another TX/RX set may specify a 9 dB power budget over communications grade fiber (50/125 graded–index fiber). These power allowances include the effects of coupling losses of TX and RX connector losses.

TX/RX specifications guarantee a maximum BER over the specified data rate range as long as the fiber meets the grade specification. A typical TX/RX set specification might be "*BER of 10^{-9} for data rates from 10 Mb/sec to 50 Mb/sec using communications grade fiber up to 2 km in length*". The BER will decrease with fiber length, so that a shorter length will exhibit an improved BER. However, manufacturers generally do not provide information on how to derate BER vs. link length or data rate. The BER will increase rapidly outside the specified operating data rate range (due to the filter characteristics of the receiver circuit).

EXAMPLE:

A TX/RX set specifies a 9–dB power budget using communications–grade fiber. If the transmitter and receiver are connected to a fiber having 2–dB/km attenuation and two connectors having 1 dB loss each, what is the maximum length of the link?

The coupling loss is included in the TX/RX specification, since the grade of fiber is specified, which in turn specifies the maximum NA.

$$P_{out} = P_s - \Sigma \text{ (losses)}$$

$$\Sigma \text{ (losses)} = P_{out} - P_s = 9 \text{ dB}$$

$$\alpha L + 2 L_c = 9 \text{ dB}$$

$$L = [\,9 \text{ dB} - 2(1 \text{ dB})\,] / \alpha = (7 \text{ dB}) / (2 \text{ dB/km}) = 3.5 \text{ km}$$

The analysis of a digital link is more complicated when the transmitter and receiver circuits are given, rather than the performance specifications. In this case, the receiver circuit must be analyzed to determine the receiver bandwidth and noise performance. Similarly, the transmitter risetime must be predicted using appropriate circuit analysis techniques. If transmitter power is not specified into a particular grade of fiber, it is necessary to determine the coupling loss. Finally, the link power and risetime budgets must both be analyzed for the selected fiber and connectors.

The advantage of knowing the TX and RX circuits is that performance tradeoffs can be made by the designer.[1] For example, the designer can readily

---

[1] The designer is referred to standard references on communications theory and on data communication techniques.

analyze the link budgets to achieve a BER of 10^{-12} rather than the more common 10^{-9}. The designer may also be able to make cost tradeoffs, such as selecting a less costly fiber for a short link. The disadvantage is that the designer must evaluate the link power and risetime budgets, rather than simply selecting the proper grade of fiber and evaluating the link power budget.

EXAMPLE:

A TX/RX set specifies a 9 dB power budget using communications grade fiber, and a data rate up to 200 Mb/sec over distances up to 1.2 km. If the transmitter and receiver are connected to a fiber having 2 dB/km attenuation and two connectors having 1 dB loss each, what is the maximum length of the link?

From the previous example, the link length limit due to power losses is 3.5 km. The link length limit due to risetime is 1.2 km, as given in the TX/RX specifications. The link length cannot exceed the shorter of these two lengths.

$L_{max} = 1.2$ km.

It may be possible to select a lower cost fiber having higher attenuation, as long as it has the required fiber bandwidth (communications–grade fiber).

8.4: Analog Links

Many analog links can be designed using modules. For example, a CCTV link can be made from CCTV TX/RX sets and communications grade fiber. As with digital modules, the TX/RX specifications will include specifications for the grade of fiber.

Analog link analysis proceeds in a similar fashion to the analysis of digital links. When TX/RX modules are selected with the appropriate grade of fiber, the analysis consists of checking the link loss against the permitted loss.

Transmitter and receiver specifications for analog links are often more detailed than for digital links. This allows the analysis to address tradeoffs between bandwidth, SNR, and link loss budgets.

EXAMPLE:

An analog link has a TX/RX set operating at 820 nm. The TX has a risetime of 3 nsec and output power of at least –2 dBm into 50/125 step-index fiber having an NA of 0.2 or greater. The TX contains an optical source whose spectral width is 40 nm.

The matching receiver has a bandwidth of 350 MHz and a sensitivity of –32 dBm. The TX/RX set is designed to be used with fiber having a bandwidth of 10 MHz–km or better.

A fiber is selected having a bandwidth of 11 MHz–km, an attenuation of 5 dB/km at 820 nm, and D = 0.12 nsec/nm/km. The link is 1 km long and contains two connectors with an average loss of 2 dB. The design margin is 4 dB. What is the bandwidth of this link?

Link power budget:

P_s	– 2 dBm	
Coupling loss (included)	0 dBm	
Attenuation (5 dB/km x 1 km)	5 dB	
Connectors (2 @ 2 dB)	4 dB	
Splices (none)		
Design margin	4 dB	
P_{out}	– 15 dBm	$\mathbf{P_{out} > P_{min}}$

Link risetime budget:

t_s	3 nsec
t_r (0.35/B)	1 nsec
Fiber	40.25 nsec
Modal (0.44 / B_0) = 40 nsec/km	
Material ($D\ \sigma_\lambda$) = 4.8 nsec/km	
t_{out}	40.4 nsec

$B_{max} = (0.35 / t_{out}) =$ 8.67 MHz

Note that for this example, the power budget must be calculated to ensure there is sufficient power to achieve the desired SNR. If there is insufficient power, then there is no need to calculate the risetime budget, since the link will not work.

EXAMPLE:

An analog link has a TX/RX set operating at 820 nm. The TX has a risetime of 3 nsec and output power of at least –2 dBm into 50/125 step–index fiber having an NA of 0.2 or greater. The TX contains an optical source whose spectral width is 40 nm. The matching receiver has a bandwidth of 350 MHz and a sensitivity of –25 dBm. The TX/RX set is designed to be used with fiber having a bandwidth of 10 MHz–km or better. A fiber is selected having a bandwidth of 11 MHz–km, an attenuation of 8 dB/km at 820 nm, and D = 0.12 nsec/nm/km. The link is 2 km long and contains two connectors with an average loss of 2 dB. The design margin is 4 dB. What is the bandwidth of this link?

Link power budget:

P_s	– 2 dBm	
Coupling loss (included)	0 dBm	
Attenuation (8 dB/km x 2 km)	16 dB	
Connectors (2 @ 2 dB)	4 dB	
Splices (none)		
Design margin	4 dB	
P_{out}	– 26 dBm	$\mathbf{P_{out} < P_{min}}$

Link risetime budget:

t_s	3 nsec
t_r (0.35/B)	1 nsec
Fiber	80.5 nsec
Modal ($0.44/B_0$) = 40 nsec/km	
Material ($D\,\sigma_\lambda$) = 4.8 nsec/km	
t_{out}	80.8 nsec

$B_{max} = (0.35 / t_{out}) = 4.33$ MHz

In this example, the power budget does not provide sufficient margin. Although the link may operate when first installed, the operating lifetime (MTTF) will be inadequate. This link will not operate according to specifications even at 1 MHz, even though the predicted bandwidth is over 4 MHz. Since the link power budget shows only a 1–dB deficit, it is likely that a more powerful transmitter or a less lossy fiber could be considered. Another alternative is a connector loss specification of 1.5 dB rather than 2 dB.

8.5: Wavelength Division Multiplexing

Wavelength division multiplexing (WDM) uses a link to carry information on more than one optical carrier wavelength. Typically, the two carriers are in different windows to maximize signal isolation, but this is not an absolute requirement. In the earliest WDM systems, a short–wavelength carrier was used in one direction, while a long–wavelength carrier was used in the other. (Because optical signals do not interfere as they travel down the link, a fiber can be used as a bi–directional transmission medium.)

It is possible to operate a WDM system with as little as 5 nm separating the carriers. When the two carriers are in separate windows, it is possible to select a pair of detectors that are sensitive only in one of the two windows, preventing detection of the other carrier. When the carriers are close together, the photodetectors will respond to any optical energy that is incident on them. In that case, additional optical filters are needed to separate the carriers to prevent light from one "link" being detected by the second.

Analysis of a WDM system proceeds in the same fashion as for a single wavelength link. The link losses must include any losses due to separating the two carrier wavelengths at each end of the link (due to filters, mirrors, etc.). The link is fully analyzed for performance for both wavelengths, and is defined to be acceptable only if it is acceptable at *both* wavelengths.

EXAMPLE:

A WDM system is designed to have an 820–nm link in one direction and a 1300–nm link in the other. It is operated over graded-index fiber having a bandwidth of 450 MHz–km. Material dispersion in 0.12 nsec/nm/km at 820 nm, and 0.05 nsec/nm/km at 1300 nm. The 820–nm source has a spectral width of 45 nm, while the 1300–nm source has a spectral width of 85 nm. Which wavelength will determine the maximum link performance?

Since both attenuation and dispersion are higher at the shorter wavelength, this link will be limited by the performance limits of the shorter wavelength. Thus the 820–nm link will determine the maximum length and the maximum bandwidth.

EXAMPLE:

A WDM system is designed to have an 850–nm link in one direction and a 1300–nm link in the other. It is operated over graded-index fiber having a bandwidth of 450 MHz–km. Material dispersion in 0.10 nsec/nm/km at 850 nm, and 0.05 nsec/nm/km at 1300 nm. The 850–nm source has a spectral width of 2 nm, while the 1300–nm source has a spectral width of 4 nm. Which wavelength will determine the maximum link performance?

Since attenuation is higher at the shorter wavelength, the maximum length will be limited by the attenuation at the shorter wavelength. Since the two sources produce equal material dispersion and the fiber bandwidth is approximately independent of wavelength, the bandwidth performance will be limited by the slower of the two sources.

8.6: Summary

Link analysis consists of evaluating the link power and link risetime budgets. When a transmitter and receiver module set is selected, together with the specified grade of fiber, the budget analysis reduces to a simple loss budget analysis.

To meet design specifications, *both* link power *and* link bandwidth budgets must be met. Evaluation of the link power and risetime budgets provides information on link limitations. The budgets sheets also provide information on whether the link is power or bandwidth limited.

8.7: Component Library for Exercises

This component library is composed of typical components, and is not representative of components from any particular manufacturer. Similarly, prices are typical (1990) values, and may increase or decrease with time. At this time, cable costs are decreasing for standard fibers and increasing for custom fibers. Transmitter and receiver costs are generally decreasing, depending on the type of transmitter and receiver selected.

LED #1: λ = 820 nm, 40–nm spectral width, 35–μm diameter, 2.3–mW output, 6–nsec risetime, $50 cost, 15–year MTTF.

LED #2: λ = 1300 nm, 70–nm spectral width, 25–μm diameter, 0.5–mW output, 2.5–nsec risetime, $500 cost, 10–year lifetime.

Laser #1: λ = 820 nm, 1–nm spectral width, 4–dB coupling loss to single–mode fiber, 0.5–mW output, 1.5–nsec risetime, $100 cost, 3–year lifetime.

Laser #2: λ = 1300 nm, 2–nm spectral width, 3–dB coupling loss to single–mode fiber, 0.1–mW output, 1–nsec risetime, $1000 cost, 2–year lifetime.

Fiber #1: 50/125 step–index, n_1 = 1.446, NA = 0.24, α = 8 dB/km @ 820 nm, α = 3 dB/km @ 1300 nm, D = 0.12 nsec/nm/km @ 820 nm, D = –0.035 nsec/nm/km @ 1300 nm, $0.50/m cost, 40–year lifetime.

Fiber #2: 50/125 graded–index, n_1 = 1.446, B_o = 500 MHz–km, NA = 0.19, α = 5 dB/km @ 820 nm, α = 2 dB/km @ 1300 nm, D = 0.12 nsec/nm/km @ 820 nm, D = –0.035 nsec/nm/km @ 1300 nm, $1.50/m cost, 40–year lifetime.

Fiber #3: 5/125 single–mode, n_1 = 1.446, NA = 0.13, α = 4 dB/km @ 820 nm, α = 0.8 dB/km @ 1300 nm, D = 0.12 nsec/nm/km @ 820 nm, D = –0.035 nsec/nm/km @ 1300 nm, waveguide dispersion = –0.05 nsec/km @ 820 nm, –0.03 nsec/km @ 1300 nm, $1.00/m cost, 40–year lifetime.

Receiver #1: Si photodiode, 5–nsec risetime, P_{min} = –35 dBm @ 820 nm and 50 MHz, derate at 6 dB/decade above 50 MHz, $100 cost, 30–year lifetime.

Receiver #2: InGaAsP photodiode, 1–nsec risetime, P_{min} = –25 dBm @ 1300 nm and 100 MHz, derate at 6 dB/decade above 100 MHz, $500 cost, 20–year lifetime.

Connector set #1: MM, 2 dB typical loss, $7.

Connector set #2: MM, 0.8 dB typical loss, $25.

Connector set #3: SM, 1 dB typical loss, $30.

8.8: Exercises

1. Why is an LED not used with single–mode fiber for most applications?
2. Why is a laser not used with multimode fiber for most applications?
3. Using Figure 8-1, determine the change in BER (in dB) when the coding scheme is changed from NRZ to RZ while the power is held constant.
4. Using Figure 8-1, determine the change in required power (in dB) when the coding scheme is changed from NRZ to RZ while the BER is held constant.
5. Complete a loss budget for a link made from LED #1, fiber #1, receiver #1, and 2 connectors #1. Use a link length of 1.0 km and system margin of 6 dB.
6. Calculate the loss budget chart for a link made from LED #2, fiber #1, receiver #2, and 2 connectors #1. Use a link length of 1 km and a system margin of 3 dB.
7. Calculate the loss budget chart for a link made from LED #2, fiber #2, receiver #2, and 2 connectors #2. Use a link length of 1 km and system marginof 6 dB.

8. Calculate the attenuation limited distance for LED #1, fiber #1, receiver #1, and 2 connectors #1. Assume a 6–dB system margin.
9. Calculate the attenuation limited distance for LED #2, fiber #1, receiver #2, and 2 connectors #1. Assume a 6–dB system margin.
10. Compare the attenuation limited distances for Problems 8 and 9.
11. Calculate the attenuation limited distance for LED #1, fiber #2, receiver #1, and 2 connectors #1. Assume a 6–dB system margin.
12. Calculate the attenuation limited distance for LED #2, fiber #2, receiver #2, and 2 connectors #1. Assume a 6–dB system margin.
13. Calculate the loss budget chart for a link made from laser #1, fiber #3, receiver #1, and 2 connectors #3. Use a link length of 1 km and a system margin of 6 dB.
14. Calculate the attenuation limited distance for laser #1, fiber #3, receiver #1, and 2 connectors #3. Assume a 6 dB system margin.
15. Complete the risetime budget chart for a link made from LED #1, fiber #1, receiver #1, and 2 connectors #1. Use a link length of 1 km and NRZ encoded data at 10 Mbit/sec.
16. Calculate the bandwidth limited distance for LED #1, fiber #1, receiver #1, and 2 connectors #1. Assume NRZ encoded data at 10 Mbit/sec.
17. Calculate the risetime budget chart for a link made from LED #2, fiber #1, receiver #2, and 2 connectors #1. Use a link length of 1 km and NRZ encoded data at 10 Mbit/sec.
18. Calculate the bandwidth limited distance for LED #2, fiber #2, receiver #2, and 2 connectors #1. Assume NRZ encoded data at 10 Mbit/sec.
19. Calculate the bandwidth limited distance for LED #2, fiber #2, receiver #2, and 2 connectors #1. Assume 4B/5B encoded data at 10 Mbit/sec.
20. Calculate the bandwidth limited distance for LED #2, fiber #1, receiver #2, and 2 connectors #1. Assume Manchester encoded data at 10 Mbit/sec.
21. Plot the maximum distance vs. bandwidth for a system, having a 3 dB design margin, made from LED #1, fiber #1, receiver #1, and 2 connectors #1.
22. Plot the maximum distance vs. bandwidth for a system, having a 3 dB design margin, made from LED #2, fiber #1, receiver #2, and 2 connectors #2.
23. Plot the maximum distance vs. bandwidth for a system, having a 3 dB design margin, made from LED #2, fiber #2, receiver #2, and 2 connectors #2.
24. Plot the maximum distance vs. bandwidth for a system, having a 3 dB design margin, made from laser #1, fiber #3, receiver #1, and 2 connectors #3.

25. Consider a system designed to last 10 years that can be made from either LED #1 or laser #1, together with receiver #1. Given the cost of replacing the sources and receivers over the life of the system, what is the total active component cost for each solution?

Chapter 9 Link Design

Link design begins with the extraction of the design constraints from a verbal description of the design problem. For example, the problem may be stated as the construction of a data network between two buildings, with taps to each individual office. Site maps would then be used to determine the physical location of each office, and the distance between the buildings.

The appropriate transmission technology[1] will depend on the distance spanned by the network and by each individual link, by the network data rate or bandwidth, and by anticipated growth needs. For example, when fiber optic cables were planned for installation in the underwater BART tunnels between San Francisco and Oakland, the highest grade of fiber available was selected. Future growth was accomplished by upgrading the transmitters and receivers for improved performance.

Future upgrades are a major consideration in planning a fiber optic network, since the cost of adding additional fiber capacity is far greater after the initial installation is completed. For example, in a fiber link under a city street, the cost of digging up the street is $200/ft, while the cost of additional single–mode fibers in a cable is around $0.10/fiber–ft. It is important to include additional fibers as appropriate, to address both breakage during installation and future growth requirements after installation. The additional fiber cost is negligible compared to the cost of later adding fiber through remodeling, since the installation cost (equipment and labor) is typically 40–70% of the total system cost.

Additional system constraints, such as system reliability, may be defined

[1] Networks may be a mix of twisted pair, coax, fiber, microwave, etc.

early in the design process or later as the system design takes shape. Over–specification of constraints may lead to excessive system cost. Often cost can be traded against reliability or redundancy. A maximum cost is often set by the cost of alternatives such as copper or microwave links.

One feature of growing networks is that a portion of the system may have been previously implemented in one technology, while growth requirements may require new technology. Either growth patterns or cost restrictions may lead to a hybrid network. A hybrid network contains a mix of technologies. In a large network, portions of the network may be implemented in wire (twisted pair and/or coax) and other portions are in fiber (multimode and/or single–mode). Typically, wire is used for the local interconnections (such as PC–printer links), while fiber is used for longer or faster sections (such as LAN connections between departments). Figure 9-1 shows such a hybrid network. The main concern in a hybrid network is to have appropriate troubleshooting tools available for the node of interest; it does no good to have a fiber test kit at an electrical wire, and vice versa.

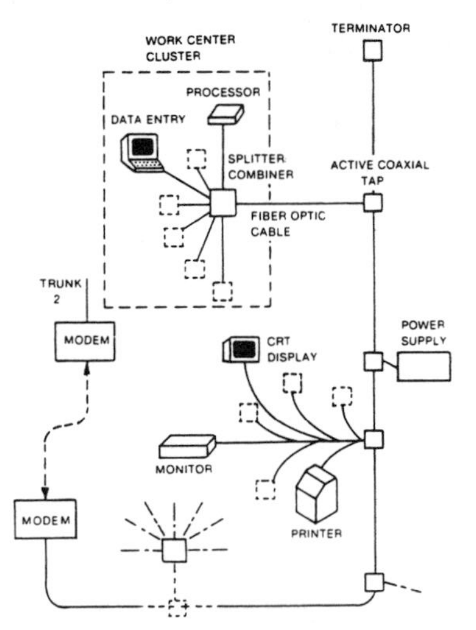

Figure 9-1: A Hybrid Network.

Courtesy of AMP Inc.

9.1: Window Selection

The first step in link design is selecting the wavelength window. The 800–nm window is suitable for low performance links, typically limited to distances under 2 km and bandwidths below 20 Mb/sec. Both 1300– and 1500–nm window operation permits operation to longer distances, with 1500–nm operation providing somewhat longer distance capability. The

1300–nm window provides optimum bandwidth with standard fiber, while the 1550–nm window provides optimum bandwidth with dispersion–shifted fiber.

There is significant improvement in performance going from the 800 nm window to the 1300–nm window, and going from multimode to single–mode fiber. There is relatively less variation in performance within a window, with the exception of the difference in performance between multimode and single–mode fibers. For some designs, a multimode link operating around 1300 nm provides comparable performance to a single–mode link operating around 820 nm.

Often the window is selected because it is called out by network standards. Normally, only one wavelength window is used within a given network. Otherwise, the designer selects the shortest wavelength window suitable to the design application. The shorter the window wavelength, the lower the link cost. If repeaters are needed, then moving to a longer wavelength or higher grade of fiber may reduce the number of repeaters and the system cost.

For a low performance system, having a link length under 50 m and a bandwidth below 5 Mbit/sec, plastic fiber may provide acceptable performance. This is the lowest cost solution, and is suitable for such applications as small office networks. For longer links or higher bandwidths, glass fiber is required. Step–index multimode fiber may be used for link B*L products up to around 10–15 MHz–km. Graded–index fibers may be used as high as 900 MHz–km, depending on the source wavelength and fiber quality. Single–mode glass fiber is used for the highest performance links.

It is possible that a system cannot be designed with a single link. In this case, the designer must select repeaters or fiber amplifiers to increase the signal amplitude. A digital repeater, or regenerator, will both amplify and re–time the signal, electronically restoring the bit signals sent from the transmitter. An analog repeater amplifies the signal electronically, without filtering or equalization. Both analog and digital repeaters add electrical noise to the signal, degrading link quality. Fiber amplifiers, on the other hand, provide direct optical amplification of the light (similar to a one–pass laser). Fiber amplifiers have a high bandwidth and low noise compared to electronic amplifiers. Fiber amplifiers reached commercial volumes in 1991, and are just beginning to show their potential.[1]

---

[1] One of the major emerging applications of fiber amplifiers is in high–count splitters, where the power division factor is high.

9.2: Module Selection

Once the wavelength window has been selected, the designer selects TX and RX modules. Each module specification sheet calls out the maximum distance and data rate. Most vendors provide matched TX and RX modules, and modules from different vendors may not be interchangeable. Inter–operability of modules is dependent on using complete standards.[1]

Modules must be selected according to the criteria for a particular design. In the case of commercial designs, such as automotive links, the main design criteria are cost and visible window operation (for ease of troubleshooting). For aerospace applications, the main design criteria are reliability and ruggedness. Often, these application–specific criteria are not spelled out in the link design specifications. The unwritten criteria are very important, and often influence the selection of the transmission technology as well as the components themselves.

Module selection will be influenced by data encoding (support circuits or functional modules). For example, different module requirements exist for analog video transmission and transmission of digitized video. If prior data encoding is performed, then it may be possible to use NRZ modules; otherwise, Manchester, FDDI (4B/5B), or other encoding may be performed in the transmitter and receiver modules. In each design, the encoding scheme in the transmitter must be matched with a corresponding decoder in the receiver. Since encoding and decoding are not standardized across vendors (or even product lines), inter–operability can be assured even if a single vendor is used.

If suitable modules are available, they will usually be less expensive than custom designed transmitter and receiver circuits. If a module set meets the requirement for the link under design, then the appropriate grade of fiber is selected and preliminary link component selection is complete.

If there is no suitable module set, then custom transmitter and receiver designs must be considered. In this case, a common rule of thumb is to estimate transmitter, receiver, and fiber risetimes to each be approximately 1/3 of the system risetime. This prevents over–specification of risetimes at an early stage of the design. The design may be made using TX and RX modules from different vendors (noting the cautions on internal encoding and decoding mentioned above), or using discrete LED or laser and photodiode components (and doing the necessary circuit design).

The LED or laser diode must be wavelength matched to the photodiode in the receiver. This is guaranteed for matched TX/RX sets. For discrete

[1] Logical (framing), electrical, modulation, and optical characteristics.

designs, the receiver sensitivity must be known at the wavelength of interest. It is necessary, but not sufficient, to ensure that the receiver photodiode operates in the same wavelength window as the source diode.

The advantage of selecting existing TX/RX sets is that the vendor performance specification incorporates all wavelength matching and internal circuit design. For a discrete design, high–speed, low–noise receiver design is difficult, and requires significant analog design experience.

The module set has several further advantages. First, modules have smaller footprint area, using less printed circuit board space. Modules result in fewer components for assembly (since all TX and RX components are internal to the module). The smaller footprint and component counts reduce printed circuit board costs. Modules also have optimized layouts, minimizing parasitic and EMI effects inside the modules, thereby optimizing bandwidth and BER or SNR. Even with modules, however, it is important to control transmission line impedance and ground bounce in the board layout.

9.3: Fiber and Cable Selection

For a module–based design, TX/RX specifications call out the grade of fiber necessary to achieve the rated bandwidth and distance. For example, a common specification is "communications grade multimode fiber", which is a graded-index fiber with certain characteristics.

A module set specification sheet will call out the maximum loss between the transmitter and receiver. The link power budget must address fiber attenuation at the operating wavelength, as well as other losses (e. g. connectors). The module specifications normally incorporate the coupling loss between the optical diode and the fiber and some power margin.

For a custom design, both the power and risetime budgets must be evaluated in detail. If the material dispersion coefficient is unknown, then the data for pure silica (SiO_2) are used as a reasonable estimate. Specification sheets for dispersion–shifted fiber will normally give plots of D vs. wavelength. For MM fiber, B_0 is most commonly given, rather than modal dispersion. For single–mode fiber, the data sheet may contain a B_0 value, which is used for waveguide dispersion calculations (since there is no modal dispersion in single–mode fiber).

A fiber data sheet will always give the profile (e. g. step–index), as well as the NA. Attenuation may be specified at one or several wavelengths. It is important to have specifications for the attenuation near the actual wavelength of interest. For example, the attenuation at 850 nm does not provide information on the attenuation at 1300 nm, since the long–wavelength attenuation is more sensitive to the OH^- content of the fiber.

EXAMPLE:

The selected TX/RX set has the following specifications:
50 Mb/sec, 2 km using communications grade MM fiber with NA > 0.20
1300–nm operation
11–dB power budget
The selected fiber has the following specifications:
Communications grade, NA = 0.22
Attenuation of 5 dB/km @ 820 nm and 2 dB/km @ 1300 nm
Attached connectors with 1.2–dB loss each
The link to be designed has the following specifications:
Length 1.5 km
50 Mb/sec
Receiver tap through a 4–dB splitter having 1–dB excess loss

Power budget:

Connector losses	2.4 dB
Attenuation loss (2 x 1.5)	3 dB
Splitting excess loss	1 dB
Splitting coupling ratio to RX	4 dB
Total losses	10.4 dB

Since the total loss is less than the specified 11 dB margin, this link meets the design specifications with this TX/RX and fiber combination.

Cable selection must take into account the number of fibers in a bundle and the environment of the installation. Cable coatings, like those of insulator coatings for wires, are selected according to their intended use. Plenum and building cables, whether wire or fiber, must meet the appropriate building codes for toxicity of any gases given off during a fire, as well as codes for fire resistance and other properties. Cable intended for direct burial in the ground must be armored for rodent protection, while cable which is laid in ducts does not need as much strength or protection.

Cables are available for a variety of uses, each with its own code and application requirements. The major classes are: duct and plenum cables for in–building use, aerial cables for stringing on telephone poles, high-temperature cables to 400°C or more for industrial applications, and direct burial cables for telecommunications and field installation.

9.4: Performance Analysis

Performance analysis begins by calculating the power and risetime budgets for the link. If the link meets all specifications, then the budgets are examined to see if lower cost components can be used. If the link does not meet specifications, then the budgets are used to identify the problem areas.

The problem areas are then addressed by selecting the appropriate components to replace the ones that cause the link to fail to meet budget. In the following example, for instance, it is the fiber that causes the link to not meet specifications. Design consists of evaluating proposed solutions (sets of components) until a set is found that meets the design criteria.

EXAMPLE:

A preliminary design results in the selection of the following components:

TX: LED, $\lambda = 850$ nm, $\sigma_\lambda = 50$ nm, P = 0.3 mW, $t_{TX} = 4$ nsec

RX: $P_{min} = -35$ dBm @ 850 nm, $t_{RX} = 5$ nsec

Fiber: Step–index, NA = 0.2, $\alpha = 6$ dB/km, $B_0 = 10$ MHz–km

2 connectors at 2 dB each, 3 splices at 0.3 dB each

Link specifications: 50 Mb/sec, 0.5 km, NRZ, 5 dB margin

Power budget:			
Source power	0.3 mW	–0.5 dBm	
Coupling loss		14 dB	
Attenuation		3 dB	
Connectors and splices		5 dB	
Margin:		5 dB	
P_{out}		–25.5 dBm	(OK)

Risetime budget:			
TX		4 nsec	
RX		5 nsec	
Fiber		22.2 nsec	
modal	22 nsec		
material	2.7 nsec		
t_{link}		23 nsec	
$t_{req} = 0.7 / 50$ Mb/sec		14 nsec	(NOT OK)

This set of components will not meet the design specifications, because it does not provide fast enough risetime. The bandwidth is limited almost entirely by the modal dispersion. The next design iteration will require either a graded–index or single–mode fiber. Since there is adequate power margin, the graded–index fiber alternative could use the same LED source. Use of a single–mode fiber would require replacing the LED with a laser source.

Once the link components have been selected, it is necessary to verify that all other design constraints are met. These include the SNR or BER, the reliability, and the cost, among other things. For a module–based design, the SNR or BER is assured as long as the module specifications are met, with appropriate derating factors applied if necessary. For a discrete design, the receiver circuits are analyzed and noise performance verified.

Reliability of a link is no better than the least reliable component. In most designs, the transmitter will fail first. The operating environment of the transmitter must be predicted, particularly the operating temperature extremes. The transmitter will usually operate at a temperature significantly higher than ambient, since it is usually in an enclosure (e. g. on a PC board inside the main PC box).

The link cost is composed of two portions: the initial component and installation cost, and the maintenance cost. The initial cost is easily calculated for the link once the components have been selected. The life cycle cost will depend on the actual reliability of the components in the field.

EXAMPLE:

The costs associated with the previous preliminary design example are: TX $250, RX $200, cabled fiber $2/ft, connectors $20 each; installation: electronics $100, fiber $15/m, connectors $8 each. Find the costs of component and installation and the total cost.

TX	$ 250
RX	$ 200
Fiber	$ 3,182
Connectors	$ 40
Component cost:	**$ 3,672**
Electronics	$ 100
Laying fiber	$ 7,500
Connectors	$ 16
Installation cost:	**$ 7,616**
Total cost:	**$ 11,288**

As can be seen in the example above, installation costs will commonly be the dominant link cost. The cost of fiber will similarly dominate the component costs. The exception to this rule is short high–speed links, where the cost of fiber may be smaller than the cost of the TX/RX set.

9.5: Summary

Link design proceeds in three steps: selecting a wavelength window, selecting a set of components within that window, and analyzing the performance of the link. The link power and risetime budgets are used for evaluating a design, and for determining the system limitations and pointing the way to improve the design.

Once a set of components has passed the budget analysis, then SNR or BER and reliability can be evaluated. The link cost depends on both the initial installed cost and on the cost of maintaining the link over its required operating life. Initial cost is usually dominated by the cost of laying the fiber, while long term costs are more sensitive to the cost of replacing the transmitter.

Link design proceeds most rapidly when a standard TX/RX set can be used for the design. Discrete designs take more design effort, and require careful attention to receiver circuit design.

9.6: Exercises

Use the same component library as in the previous chapter to perform systems design. For each design, show that both power budget and risetime budgets are met at the lowest link cost.

1. Why is LED #1 used with receiver #1 and not receiver #2?
2. Why is LED #2 used with receiver #2 and not receiver #1?
3. Why can fiber #1 be used with either LED transmitter?
4. Why should fiber #3 not be used with either LED transmitter? Why isn't fiber #1 normally used with the laser transmitter?
5. Assuming a 6 dB design margin, one connector at each end of the link and a repeater made from LED #1, receiver #1, and two connectors #1, make a plot of cost vs. link length for a 10 MHz link made from LED #1, receiver #1, and fiber #1, for link lengths from 0 to 5 km.
6. Assuming a 6 dB design margin, one connector at each end of the link and a repeater made from LED #1, receiver #1, and two connectors #1, make a plot of cost vs. link length for a 10 MHz link made from LED #1, receiver #1, and fiber #2, for link lengths from 0 to 5 km.
7. Assuming a 6 dB design margin, one connector at each end of the link and a repeater made from LED #2, receiver #2, and two connectors #2, make a plot of cost vs. link length for a 10 MHz link made from LED #2, receiver #2, and fiber #1, for link lengths from 0 to 5 km.
8. Assuming a 6 dB design margin, one connector at each end of the link and a repeater made from LED #2, receiver #2, and two connectors #2, make a plot of cost vs. link length for a 100–MHz link made from LED #2, receiver #2, and fiber #2, for link lengths from 0 to 5 km.

9. Compare the costs of a 2–km, 5–MHz link designed from (a) LED #1, receiver #1, two connectors #1, and fiber #1; (b) LED #1, receiver #1, two connectors #1, and fiber #2; (c) LED #2, receiver #2, two connectors #2, and fiber #2, and (d) laser #1, receiver #1, two connectors #3, and fiber #3. Assume a 6–dB design margin.
10. Compare the costs of a 1–km, 50–MHz link designed from (a) LED #1, receiver #1, two connectors #2, and fiber #2; (b) LED #2, receiver #2, two connectors #2, and fiber #2; (c) laser #1, receiver #1, two connectors #3, and fiber #3; and (d) laser #2, receiver #2, two connectors #3, and fiber #3. Assume a 6–dB design margin.
11. Design a repeaterless 2–km link, operating at 10 Mbit/sec with Manchester-encoded data, and having a 10–dB system margin. Show all loss budget, risetime budget, and cost budget calculations.
12. Design a repeaterless 0.5–km link, operating at 10 Mbit/sec with NRZ-encoded data, and having a 10–dB system margin. Show all loss budget, risetime budget, and cost budget calculations.
13. Design a repeaterless 5–km link, operating at 1 Mbit/sec with NRZ-encoded data, and having a 10–dB system margin. Show all loss budget, risetime budget, and cost budget calculations.
14. Design a repeaterless 2–km link, operating at 25 Mbit/sec with 4B/5B-encoded data, and having a 10–dB system margin. Show all loss budget, risetime budget, and cost budget calculations.

Chapter 10 Installation and Testing

The actual performance of an installed link is frequently different from the designed performance. Initially, there is (or at least should be) a larger link margin than there will be at the end of life. There will also be some excess bandwidth. It is also possible for an installed link to have less performance than intended, in which case the link will not meet specifications. This may occur, for example, due to bending losses greater than allowed for in the power budget.

In one case history, a link was observed to function under only limited conditions. The bandwidth specification was 350 MHz, so the designers selected a 400–MHz receiver amplifier. This did not leave sufficient worst-case risetime budget for the transmitter risetime and fiber dispersion. The link was functional only after days of manual "tuning" of connections and transmitter current modulation. While such a link may be acceptable for functional demonstrations, it is not suitable for reliable network operation.

It is important that both components and the installed link undergo testing to verify functional operation. Additional testing may be performed over the life of the system to ensure continued functional operation. For example, a digital link may report data errors (e. g. parity errors) to the system. A large number of errors will alert the system manager to loss of link quality.

10.1: Component Acceptance Testing

Component testing is performed prior to installation of the components into a link. There are two basic test goals: power acceptance, and performance acceptance. Power acceptance testing is straightforward, and is normally performed on all components. Performance acceptance testing requires more extensive tests, and may not be required for all components.

Acceptance testing of functional module sets begins with power testing of the transmitter. Functional sets have already passed testing of functional specifications by the vendor, and acceptance testing is mainly to verify continued operation. Functional testing is done by attaching the module set into a reference link (standard fiber and connectors), and verifying data rates and bit error rates. This functional test verifies the power margin and data quality for the TX/RX set.

Analog components require testing of the risetime, or bandwidth, and the SNR in addition to the optical power. Analog tests are more sensitive to the electronics of the reference link (particularly for high–speed links, which are very sensitive to such things as impedance matching). Analog testing is also sensitive to the electrical circuit effects of the test equipment: a link may be functional with the test equipment in place, but not when it is removed (or vice versa). This is because the test equipment adds parasitic capacitance to the test point, loading the link electronics. The finite bandwidth of test sources and instruments also influences the observed bandwidth. For example, it is impossible to test a 500–MHz link with a 100–MHz scope, since the oscilloscope risetime will be much slower than the link risetime, dominating the measured risetime.

Additional tests may be made on the module set. Testing for temperature effects, particularly for laser transmitters, may be required. Transmitter reliability may be checked by testing some transmitters to failure. Vendors may test for parameters such as source spectral width and reliability failure mechanisms, but these are generally not part of the component acceptance test suite.

Fiber acceptance testing also begins with power testing. A known power is launched into the fiber using a reference source, and the output power is measured. This test, performed on an entire reel of fiber, verifies that the fiber meets attenuation requirements. Functional testing is done by attaching the fiber into a reference link (standard TX, RX, and connectors), and verifying data rates and bit error rates. This functional test verifies the power budget (coupling and attenuation effects) and bandwidth for the fiber. The reference link operational wavelength is generally the same as the operational link wavelength to ensure correlation of reference link performance to installed link performance. Acceptance tests must be performed on each fiber within a cable.

Fiber attenuation and power transmission depends strongly on the mode distribution within the fiber. For a short fiber, there is a significant portion of the power in the higher order modes, while for a longer fiber the power is more concentrated in lower order modes. A short length of fiber may be used to emulate a long fiber by use of a "mode stripper". The mode stripper introduces several small bends into the fiber, which strips out the outer modes. This has the advantage of allowing a short length of fiber to be used for testing. There is a risk in this approach, in that stripping out too many modes will lead to excessively high performance predictions, while stripping out too few will lead to excessively low ones. Mode stripping should be used only where the performance of the tested fiber can be correlated through field data against the actual fiber performance.

Fiber may undergo a number of other tests. Bending loss characteristics may be measured using the reference link and a variable controlled bend. Fiber pull strength and breaking radius may be tested on a short section. Cable pull strength may be tested prior to pulling the cable into ducts. Reliability testing may include humidity resistance, pull strength after prolonged tension, and repeated bend strength.

Connectors are generally not tested until after they have been installed on a fiber. The primary test for connectors is power transmission loss. Connectors may also be tested for repeatability of loss characteristics over many connect/disconnect cycles. Similarly, splitters, combiners, and other passive components are tested for power transmission characteristics, as are fiber amplifiers.

10.2: Link Acceptance Testing

Link testing is an integral part of the installation process. Each point–to–point link within a network must be individually tested. Each link is tested for end–to–end power transmission, using a calibrated transmitter and receiver. This test verifies the link power margin, including the effects of bends, splices, and other losses introduced during installation. Verification includes comparing the measured losses to the link loss budget, and providing information back to the designers.

Link performance testing is done to verify the transmission (data) quality. The BER is measured at the full data rate if possible. However, for very high speed links, there may not be a BER tester in existence. In this case, the BER is measured as a function of received power, using a calibrated optical attenuator. This data is plotted, and the BER at the actual link output power is extrapolated.

Power (continuity) testing must be done on each individual fiber during installation. Bandwidth testing is done on the installed link, and BER is verified. After installation, it is desirable to monitor either transmitter power (through back facet monitors) or receiver power, since a decrease in transmitter output over time will eventually lead to link failure. By monitoring the power, the transmitter replacement can be done as scheduled maintenance, preventing an unexpected link failure.

10.3: Test Equipment

Optical component testing uses calibrated power sources and optical power meters. Sources are usually lasers, although LED sources are also available. Calibrated sources operate at a single wavelength, commonly 820, 850, 1300, or 1550 nm. Optical power meters are calibrated at the same wavelength as the source. All calibrated components should be traceable back to National Institute of Standards and Technology (formerly the National Bureau of Standards).

The most commonly used piece of test equipment is the optical power meter. An optical power meter consists of a photodiode and amplifier, usually with a digital readout. The amplifier is designed for linearity and wide dynamic range. Optical power meters can be calibrated at more than one wavelength within either the short– or long–wavelength window. Figure 10-1 shows the block diagram of a typical power meter. Power meters are designed for low noise, and typically have very low bandwidth. Power meters may also include signal averaging to improve noise rejection.

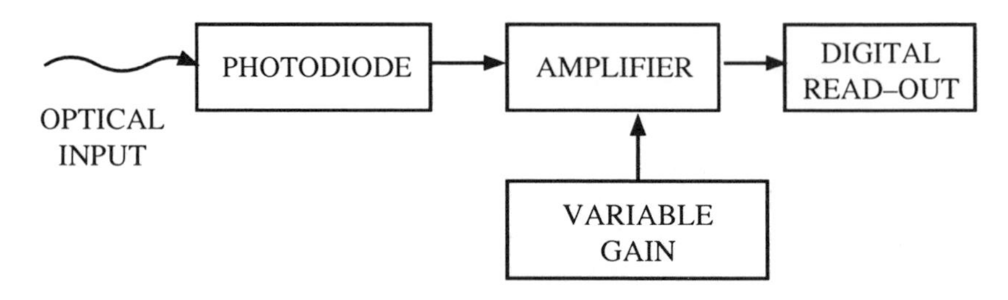

Figure 10-1: Power Meter Block Diagram.

Optical link testing is performed using an optical time domain reflectometer (OTDR). This instrument launches optical pulses into the fiber, and measures reflected power. Power is reflected by Rayleigh scattering in the fiber, and at splices and connectors due to changes in fiber index. When the index of refraction is known, the reflection time provides the distance to the reflection point. In this way, connectors, splices, and fiber faults can be located. Figure 10-2 shows an OTDR block diagram, and Figure 10-3 shows a typical OTDR display. A variety of OTDR instruments are available, with distance resolutions from a few millimeters to several meters, and ranges from 1 kilometer to tens of kilometers. Usually there is a tradeoff between resolution and range. At least one vendor has an OTDR without the graphics display, intended mainly for fault location.

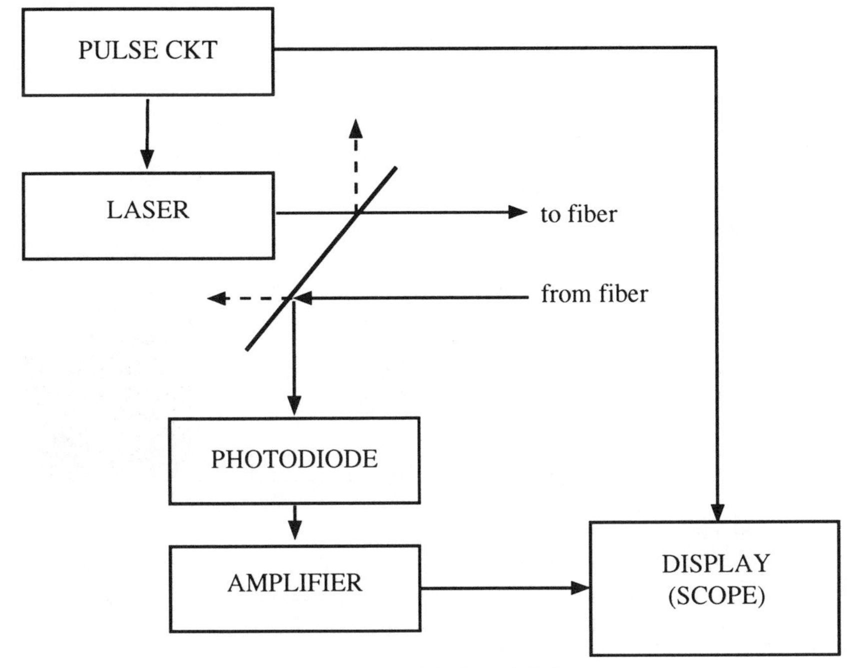

Figure 10-2: OTDR Block Diagram.

Since fiber attenuation is a function of power distribution in the modes, an OTDR ideally should use the same type of source as well as the same wavelength as the link. An OTDR normally comes with a laser source. Source modules may be available to allow selection of source wavelength and LED or laser options; these must be specified at the time the OTDR modules are purchased. In general, an OTDR has many of the options of high–quality oscilloscopes, including calibration adjustments (for example, calibrating time into distance by dialing in the index of refraction). There may also be printer or floppy drive attachments for permanent recording of the output.

The system used to produce the display shown in Figure 10-3 is an ANTEL OTDR with the analysis software and optical modules integrated into a personal computer. The use of a PC allows the user to use standard PC drives and modems to transfer the data and to save it for later use. In later tests, the link performance can be compared to the original performance. A trend is developing of instrument makers moving away from proprietary microprocessors into PCs with customized applications software.

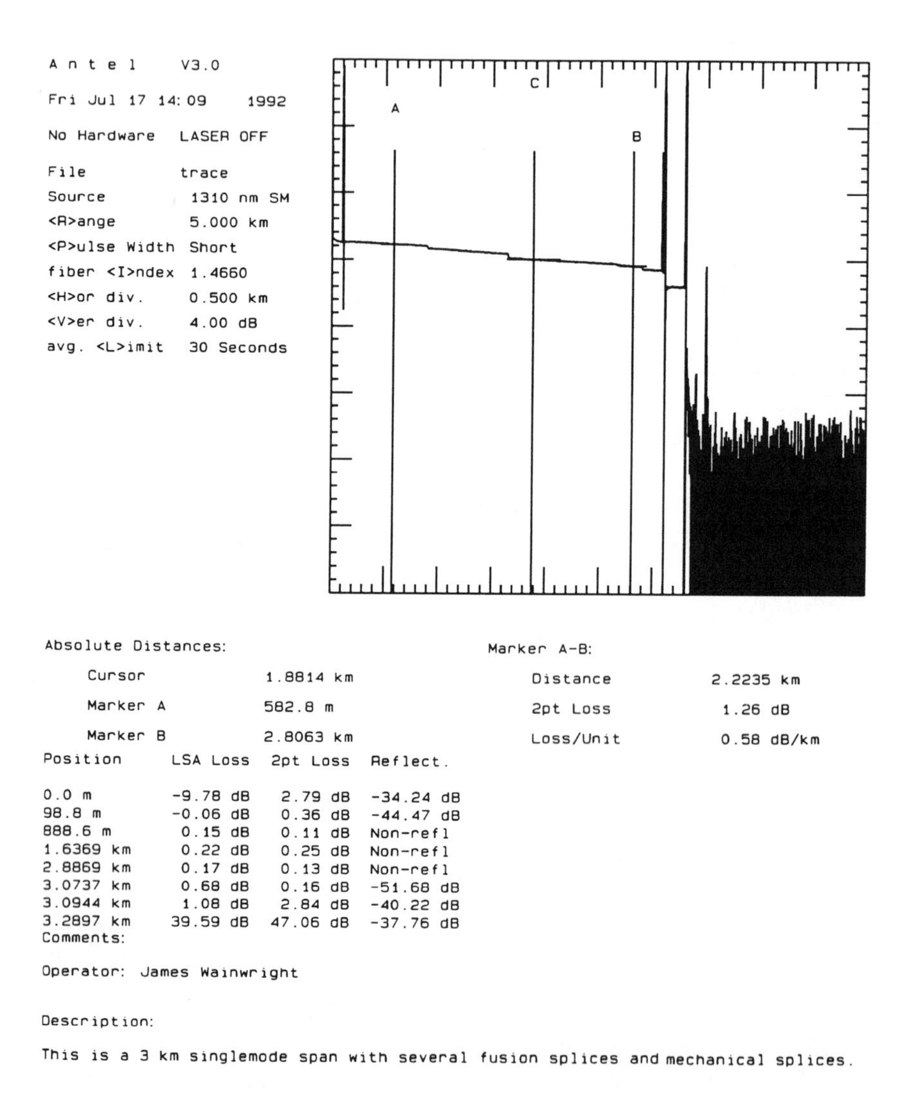

Figure 10-3: OTDR Display.

Courtesy of ANTEL Optronics Inc. and of BERT Technologies.

Both power meters and OTDRs have multiple wavelength and multiple window options. Multiple window options are achieved by providing a separate plug–in transmitter and receiver module for each wavelength window.

10.4: Link Reliability

The link MTTF is related to the MTTF of the individual components and to installation and maintenance practices. The MTTF of the link due to component failure is given by:

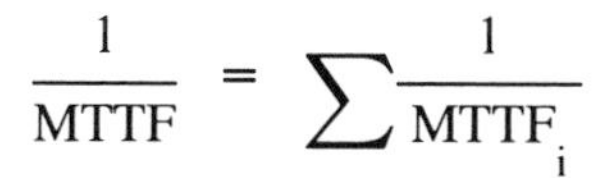

$$\frac{1}{MTTF} = \sum \frac{1}{MTTF_i}$$

where each term in the sum is the MTTF of one component in the link (transmitter, receiver, fiber, connector, etc.)

As can be seen from the equation, the shortest component MTTF will dominate the link MTTF. In the typical link, the optical source is the most likely to fail, since the optical output power degrades with time. For this reason, monitoring the output power of the source and replacing it before the link fails is an important part of regular maintenance. The second most common source of failure is in the electronics, and in the connectors during connect / disconnect operations.

Another major cause of link failures is related to the handling of the fiber and connectors. Left alone, these components have MTTFs of decades. However, frequent handling of connectors can lead to bend fatigue of the fiber. Frequently, inexperienced users bend the fiber past its breaking point as connectors are connected or removed from a link. Individual fibers may snap during installation, particularly if bend radius and pull tension are not closely monitored. Fibers may break after installation due to inadvertent acts, such as dropping an object on the cable.

The most common failure mode for fiber in the telecommunications industry is so–called "backhoe fade", where the cable is actually severed by a backhoe or other heavy equipment. West of the Mississippi, fiber breakage may be caused by rodents, which chew through the plastic protective layers; this is why most direct burial cables include armored sheaths. The link failure rate due to handling is dependent on the location of the cable runs (in–building/ underground/aerial installation), and on the number of connect/disconnect cycles. Prediction of the link failure rate (e. g. for life cost estimates) depends being able to predict the reliability of the fiber and components under actual installed conditions.

10.5: Installation Techniques

Installation of fiber optic electronics requires the skills associated with similar electronic assemblies. Perhaps the greatest concern is with electrostatic discharge (ESD) protection. ESD can cause the photodiode dark current to increase, which can seriously degrade SNR. The degradation will not necessarily cause the link to fail its power tests, but it will show up in BER testing.

Care is required in the installation of cables due to the ease with which glass fibers can be broken. Cable pull strengths depend on the cable structure (e. g. Kevlar or metal strength members). The pull strength of an individual fiber is relatively low, comparable to fishing line. When pulling cable through ducts, the tension should be monitored with a tension gauge to ensure the cable pull strength is not exceeded. Even with the best of care, the odds are high that at least one fiber will break, which is why cables are usually selected with at least two extra fibers (except for in–duplex links).

Fiber exhibits very poor bend resistance. A typical fiber will break at a bend radius around 1–2 cm. The breakage radius is given on fiber specification sheets. During installation, all sharp edges should be kept away from the fiber, since placing the fiber under tension will snap it as it crosses that edge. Thus, for example, a fiber should never be gripped with a pair of pliers, which have a sharp edge compared to the 1 cm bend breakage radius. Similarly, wrapping the cable around your hand to get a better grip means that you may end up with a better grip on a broken cable.

Connectors are another point of failure. The connector housing provides a sharp edge by its very nature. Bending the fiber, even through so simple an action as bumping it near outside of the connector, can be enough to cause a break. Strain relief sleeves should be installed on all connectors that will see frequent connect/disconnect cycles, or handling by end users (such as computer operators).

A loss of alignment or an increase in reflection loss at a connector may cause the link to fail its power budget. Connectors must be installed with proper alignment between the optical source and the fiber core. Some connectors also require index matching fluids or gels to be installed. Keyed connectors must be aligned so the two sides of the "key" are aligned. During installation, connectors may be aligned using optical transmission measurements (with a laser and power meter) to optimize alignment.

Splice installation can be done using either mechanical or fusion splices. Both types of splices can be improved using optical transmission measurements to check the transmission loss and thereby the alignment of the two fiber cores. After completing a splice, it must be properly enclosed to ensure continued resistance to humidity and other environmental effects. The strength of the completed splice and its fixture is greater than that of the fibers. However, the splice fixture again provides a sharp edge, and has the same bend breakage problem as does a connector.

Because the glass is thin (125 μm) and fragile, a fiber optic cable must be handled more carefully than a copper cable. The basic rules for cable installation and handling are summarized in Table 10-1. Most of these rules deal with the prevention of fiber breakage due to excessive bending or pulling.

Table 10-1: Cable Installation Rules

- Always handle fiber as gently as possible.
- Never pull directly on a fiber. Pull on Kevlar or a strength member.
- Never use pliers or tweezers to get a grip on the cable.
- Never twist the cable around your hand or fingers.
- Always keep cable ties loosely snugged.
- Never pull or tug on the connectors.
- Never bend or pull the fiber at the connector or splice.
- Always keep bends larger than 4 cm (1.5 in).
- Always pull long runs with a tension monitor.
- Never leave a fiber under tension if it can be avoided.
- Always leave any cable coils 30 cm (12 in) or more across.
- Always lay cables in raceways; do not pull them.
- Never shake or snap the cable loose if it sticks.
- Never crush, snag, crimp, or drop things on a cable.
- Always use an appropriate lubricant when pulling cables in ducts.
- Never walk on the cable; always step over it.
- Always keep protective caps on connectors when not in actual use.

10.6: Troubleshooting Techniques

Link failure may be observed as an increase in BER, or as a total failure in data transmission. Total failure in data transmission may be a result of a broken fiber, a source with no optical output, or a laser that no longer reaches threshold. Fiber breakage does not cause intermittent failures, unlike wire breakage. Total failures may also be caused by circuit failures (for example, loss of modulation).

A degradation of BER can be detected by decoder error detection. For example, an error detector might count parity errors, and report error rate counts above a certain minimum. The BER may be monitored by the network, and high BER reported to the system manager software. If no monitoring is performed or monitored outputs are not checked, then the first indication of failure may be an irate user.

A BER problem is most often caused by a decrease in transmitter optical power. Many transmitters include a source power monitor (e. g. using back facet monitoring of the laser diode). A receiver module may contain a "received power" monitor, with a "low power" indicator. These monitors are checked as the first step in troubleshooting a link exhibiting a high BER or a total loss of data transfer.

Both transmitter optical power and received power can be checked with a power meter. The power meter provides more information than a simple indicator signal. If the power out of the transmitter is low, then electrical connections to the transmitter should be checked (particularly external resistors and power supply voltages). If the electrical connections are OK, the transmitter should be replaced.

If the transmitted power is OK but the received power is low, then connectors should be cleaned; a small speck of dust can cause a high optical loss. If the received power is still low, then the fiber should be checked for bending, such as that caused by laying a heavy object over a cable. If the received power is zero, then the fiber should be checked for breakage.

An OTDR can be used to locate fiber faults from either end of a link, since the OTDR contains its own transmitter and receiver. To improve distance resolution, it is desirable to use the OTDR from the end nearest the break. The OTDR can also be used to check connector and splice losses. These measurements, when compared to the link loss budget, are useful in troubleshooting. The OTDR is useful both for those links with no power arriving at the receiver and those with insufficient received power.

If both the transmitter and receiver power are OK, then the modulation should be checked. Transmitter electrical failures may cause a loss of modulation without a corresponding loss of optical power. Modulation is checked by using a functional receiver to check the output for modulation. This can be done at either the transmitter or receiver end of the link.

When a link contains repeaters, then the above steps must be taken for each fiber link between electronic repeaters. For optical amplifiers (direct optical repeaters), testing is done at the transmitter and receiver. Since no light travels past a break, an OTDR will find only the nearest fiber break or the nearest electronic repeater.

Identification of active (signal carrying) fibers can be performed while a link is carrying data. Data can be tapped from a fiber by introducing a small bend. A small amount of light escapes from the fiber (bending loss), and a properly placed receiver can detect this light. This technique is useful for identifying active and dark (unused) fibers within a cable. If the characteristics of the data are different for different active fibers, then the active fibers can also be uniquely identified.

Dark fibers are useful during troubleshooting, as they provide a communications link between the two ends of the fiber. Optical talk sets, similar to telephone sets, can be connected to the dark fiber, allowing the people at the two ends of the link to exchange information. This is particularly useful during link loss measurements, where one person has the calibrated transmitting source and the other has the power meter at the other end of the link, perhaps several kilometers away.

10.7: Summary

Component testing tests DC power characteristics for all components. Data transmission characteristics are tested for the transmitter and receiver, and fiber bandwidth tests may also be made. Acceptance tests use calibrated optical sources and power meters.

Link testing measures power transmission and losses for comparison to the link loss budget. Data quality is checked at the full data rate if possible, otherwise the BER is extrapolated from a plot of the BER vs. received power.

Link failures due to fiber breakage appear as "open circuits", with no data transfer. Fiber breakage does not cause intermittent failures, unlike coax cable failures. Data loss may also be caused by a laser failing to reach threshold, by total loss of optical output power, and by circuit failures.

10.8: Exercises

1. Define the following terms: calibrated transmitter, OTDR, and dark fiber.
2. Define the following terms: attenuator, optical power meter, and active fiber.
3. Explain why the display of optical power vs. distance is a straight line for a fiber segment on the OTDR image shown in Figure 10-3.
4. A fiber link is observed to have a sudden change in the BER from 10^{-9} to 10^{-4}.
   (a) What is the first component to examine?
   (b) What other components should be checked if the first is OK?
5. A link is observed to have a sudden (less than 5 minutes) change in received power from –30 to –50 dBm. List 3 possible causes.
6. A link has a slow (over several days) change in received power from –30 to –50 dBm. What is the most probable cause?
7. Two 50/125 step–index fibers are joined by a connector. Assuming a circular particle of dust is caught in the connector, what loss will result if the particle's diameter is (a) 5 μm, (b) 25 μm, (c) 50 μm. (Hint: transmission is proportional to the unblocked core area.

Appendix A Bibliography

There are many books and magazines covering fiber optics, from components to systems to applications. Fiber optic articles also appear in many LAN and computer magazines. Some that are of interest to those working with fiber optics include the following, many of which are available in technical libraries:

Magazines And Journals:

Applied Optics
Bell System Technical Journal
IEEE Circuits and Systems Magazine
IEEE Journal of Lightwave Technology
IEEE Lightwave Communications Systems Magazine
IEEE Lightwave Telecommunications Systems Magazine
IEEE Photonics Technology Letters
IEEE Spectrum
Proceedings of the IEEE
Scientific American
Fiber Optic Product News
Laser Focus World
Lasers and Optronics
Lightwave
Photonics Spectra
Sensors

Books:

Donald Baker, *Monomode Fiber–Optic Design*, Van Nostrand Reinhold, 1987.

William Beyda, *Basic Data Communications*, Prentice Hall, 1989.

Brian Culshaw and John Dakin, eds., *Optical Fiber Sensors Vol. 2: Systems and Applications*, Artech House, 1989.

John Dakin and Brian Culshaw, eds., *Optical Fiber Sensors Vol. 1: Principles and Components*, Artech House, 1988.

Chris Gahan, *FDDI Business Communications for the 1990s*, BICC Data Networks Inc., 1990.

John Gowar, *Optical Communication Systems*, Prentice Hall, 1984.

IBM 3044 Fiber–Optic Channel Extender Link: Fiber Cable Planning and Installation Guide (1987), Publication # GC22–7073–1, International Business Machines Inc.

Gerd Keiser, *Optical Fiber Communications*, 2nd edition, McGraw Hill, 1991.

Karl Kummerle, John Limb, and Fouad Tobagi, *Advances in Local Area Networks*, IEEE Press, 1987.

C. D. Motenbacher and F. C. Fitchen, *Low–Noise Electronic Design*, Wiley–Interscience, 1973.

Peter Runge and Patrick Trichatta, eds., *Undersea Lightwave Communications*, IEEE Press, 1986.

William Stallings, *Local Networks: An Introduction*, MacMillan Press, 1984.

Andrew Tannenbaum, *Computer Networks*, Prentice Hall, 1981.

Van Nostrand's Scientific Encyclopedia.

Amnon Yariv, *Optical Electronics*, Holt Reinhart and Winston, 1985.

Appendix B Glossary

Al	chemical symbol for aluminum
As	chemical symbol for arsenic
APD	avalanche photodiode
balanced	signaling code with an equal number of high and low states
BER	bit error rate
Ga	chemical symbol for gallium
Ge	chemical symbol for germanium
ILD	injection laser diode
In	chemical symbol for indium
IR	infrared
LD	laser diode
LED	light–emitting diode
Manchester	balanced signaling code, used at lower data rates
NRZ	on–off signaling code
P	chemical symbol for phosphorous
PIN	positive–intrinsic–negative photodiode
Quaternary	made from four different elements
Repeater	analog amplifier, returns signal to original strength
Regenerator	digital amplifier, both amplifies and restores edges
RX	receiver
RZ	signaling code
Si	chemical symbol for silicon
SNR, S/N	signal–to–noise ratio (often given in dB)
Ternary	made from three different elements (e. g. AlGaAs)
TX	transmitter
UV	ultraviolet
4B/5B	balanced signaling code, used in FDDI

Appendix C Symbols, Constants, and Conversions

It is important to note that a symbol may represent any of several variables. For example, the symbol **L** may represent loss, link length, or inductance. Often the intended representation is clear from the context; dimensional analysis can be used for verification. ***In no case should a value be substituted for a symbol until the intended meaning is clear.***

The symbols presented in the tables are of three types: constants, variables, and variable constants. An example of a constant is the speed of light in free space: true constants never change their values. An example of a variable is the link SNR: it decreases over time as the link performance degrades. An example of a variable constant is the optical frequency of an LED: while the wavelength is essentially constant for any one LED, selecting a different LED may change the wavelength.

Which constants are truly variables and which remain fixed for a given design problem will depend on the specific situation. For example, the maximum link length will vary with the selection of components, while the desired link length is generally fixed by building location or other physical constraints. (It is generally difficult to move buildings to accommodate designers.)

The designer should be clear from the outset as to the classification of category for each variable. Units in the following table are given in MKSA (International System) units or the units in common usage within the optical communications industry.

Table C-1: Constants and Conversions

Constants:

$c = 2.998 \times 10^{8}$ m/sec	speed of light
$h = 6.626 \times 10^{-34}$ Joule–sec	Planck's constant
$hc = 1.240$ eV–μm	
$q = 1.602 \times 10^{-19}$ Coulomb	magnitude of electron charge
$k = 1.381 \times 10^{-23}$ Joule / Kelvin	Boltzmann's constant
$kT / q = 0.026$ eV	at room temperature (300 Kelvin)
$\varepsilon_0 = 8.854 \times 10^{-12}$ Farad / m	permittivity of free space
$\mu_0 = 4\pi \times 10^{-7}$ Henry / m	permeability of free space
$m_e = 9.11 \times 10^{-31}$ kg	mass of electron

Dimensional conversions:

1 μm = 1 micron = 1×10^{-6}m = 1×10^{-4}cm

1 eV = 1.602×10^{-19} Joule

π radians = 180°

T (Kelvin) = T (^{o}C) + 273

1 meter = 3.281 ft.

1 km = 0.6214 mile

1 mile = 1.609 km

1 mil = 10^{-3} in

1 ft = 12 in

Table C-2: Symbols and Definitions

Symbol	Units (MKSA)	Represents
a	meters (μm)	fiber radius (half of diameter)
B	bit/sec (Mbit/sec)	digital bandwidth, data rate
B	Hz (MHz)	analog bandwidth
B	Tesla	magnetic field
B_o	MHz–km	fiber distance–bandwidth product
c	m/sec	speed of light *in free space*
C	Fd	capacitance
C_A	Fd	amplifier input capacitance
C_d	Fd	diode capacitance
d	meters (μm)	offset between two fibers
dB	dimensionless	power ratio, relative power
dBm	dimensionless	power relative to 1 mW
dBu	dimensionless	power relative to 1 μW
dBn	dimensionless	power relative to 1 nW
dBW	dimensionless	power relative to 1 W
D	nsec / nm / km	fiber dispersion coefficient
DR	1 / sec	data rate
E	volts / m	electric field
f	Hz	frequency
h	Joule–sec	Planck's constant
i	Amp (nA)	modulation current
I	Amp (nA)	DC or bias current
i_n	Amp (nA)	noise current
I_s	Amp	junction saturation current
I_{th}	Amp (mA)	threshold current
k	Joules / Kelvin	Boltzmann's constant
k	1 / m	wave number, propagation constant
L	dB	loss
L	km	link length, distance
L	Henry	inductance
M	dimensionless	number of modes
M	dimensionless	photodiode gain

Symbol	Units (MKSA)	Represents
n	dimensionless	diode ideality factor
n	dimensionless	index of refraction
NA	dimensionless	numerical aperture
P_c	Watts	fiber–coupled optical power
P_s	Watts	source optical power
P_o	Watts	output optical power
q	Coulomb	*magnitude* of electron charge
R	ohms	resistance
R_A	ohms	amplifier input resistance
R_b	ohms	bias resistance
R_f	ohms	feedback resistance
s	meters (μm)	separation of two fibers
t	sec (nsec)	time
t	meters	thickness
T	Kelvin, centigrade	temperature
T_b	sec (nsec)	bit period
T_{trip}	sec (msec)	trip delay time
v	m / sec	speed of light in a material
v_d	m / sec	carrier drift velocity
V	dimensionless	V-parameter
V_{br}	volts	breakdown voltage
V_{cc}	volts	bias voltage
V_t	volts	thermal voltage (k T / q)
α	dB / km	attenuation
ε	Farad / meter	permittivity
ε_r	dimensionless	relative permittivity
λ	meters (μm)	wavelength
μ	Henry / meter	permeability
μ_r	dimensionless	relative permeability
σ_λ	meters (nm)	spectral width
τ	sec	time constant
τ_r	sec (nsec)	risetime
θ	degrees, radians	angle

Symbol	Units (MKSA)	Represents
θ_c	degrees, radians	critical angle
ω	radians / sec	angular frequency ($2 \pi f$)
Ω	steradians	solid angle

Appendix D Signals and Noise

The detection of signals in the presence of noise is critical in fiber optic systems. Because the signal undergoes significant attenuation and pulse broadening between the transmitter and receiver, it is received as a small signal.

A signal is characterized by its frequency content. The bandwidth is determined by the bandwidth of the original information source and by the modulation technique. In analog modulation, the signal may be FM modulated, or the information may be encoded on the amplitude of the optical carrier. In digital modulation, the most common technique is to switch the optical amplitude between one of two levels ("high" and "low").

Noise is characterized by its power (RMS value) over bandwidth. Most noise has a bathtub curve characteristic: flat over much of its bandwidth, rising at both high and low frequencies. The lower and upper corner frequencies are a function of the components. For example, MOSFETs often have a lower corner frequency in the audio range (1 to 10 kHz), Below this corner frequency, the power rises as 1/f, causing the noise to be called 1/f noise or "pink" noise. In the midband, the noise is referred to as "white noise", since it is uniform over frequency.

Noise is characterized further by its average and RMS values. Noise has a zero average value; that is, on the average it will be above the signal as often as it is below the signal. For low frequency measurements, noise can be removed by averaging many measurements. Unfortunately, this is not feasible with high–speed communication systems.

The impact of noise on a link can be measured in terms of the signal–to–noise ratio (SNR) or bit error rate (BER). The SNR will depend on the modulation scheme, the average optical power, and the noise. In most links, the noise is dominated by the receiver.

If one assumes a single frequency of modulation, then the received power can be written as:

$$P_r(\mathrm{t}) = (1 + m \sin(\omega \mathrm{t})) \; P_o$$

P_o is the average (DC) optical power at the receiver, P_r is the instantaneous value, and m is the modulation index. Note that Fourier analysis allows this single frequency to be expanded to address the SNR over all frequencies.

The noise in the receiver will depend on the power received. Because the contribution of the quantum noise is dependent on the number of photons, the noise will depend on the modulation index (m). Similarly, for a digital system, the noise will depend on whether the power at time t represents a “1” or a “0”; “1” is represented by a higher power level, and therefore has a higher quantum noise.

Because the quantum noise contribution is dependent on the data bit (“1” or “0”) being received, the average noise will also depend on the average number of ones and zeros in the received optical power. A code having an equal number of highs and lows is referred to as a “balanced” code. For example, in Manchester coding, the fraction of high and low states is always equal, and it is a balanced code. RZ is an unbalanced code, while NRZ has the same balance as the input data.

The number of errors in a received data stream will also depend on how the threshold is set. Figure D-1 shows a typical eye pattern outline, with a threshold detection level superimposed. All signals above the threshold are interpreted as a “1”, while all signals below are interpreted as a “0”. If the threshold is shifted, the number of bits that are interpreted as a “1” will change, even though the original data stream is unchanged. This illustrates how sensitive the BER is to the threshold level.

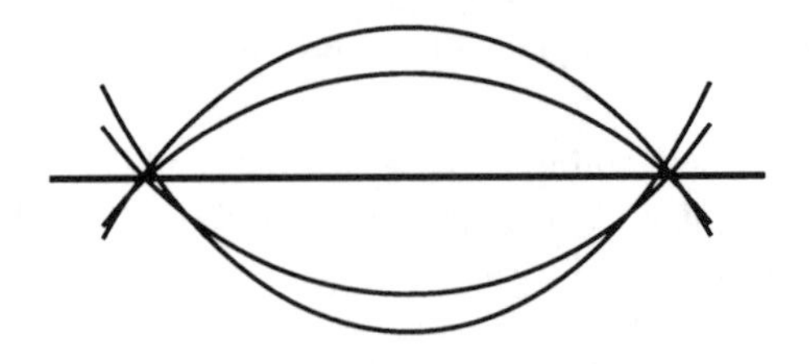

Figure D-1: Eye Pattern with Threshold Level.

As the transmitter power degrades with time, the BER will also degrade unless the threshold level is decreased. Many receivers use an active threshold control to keep the threshold level midway between the high and low signal states. This keeps the threshold near the center of the eye pattern. This can be checked using a (high–speed) oscilloscope for links where an analog output pin is available.

To perform performance predictions of the BER, it is necessary to make some assumptions about the characteristics of the noise and the signal. The most common assumptions are that the noise has been filtered to have the same effective bandwidth of the signal, and the noise has a Gaussian distribution with an average value of zero, and the same RMS value in both the high and low state. Note that while the assumption of equal RMS values is technically violated in actual receiver systems, it is adequate for performance prediction when quantum noise is a small contribution to the total noise.

Under these assumptions, the probability of error becomes the probability of a "0" having a total received power above threshold and of a "1" having a total received power below threshold. The probability of error is P(E), where a and b are the number of 1's and 0's respectively in the data stream, is:

$$P(E) = a\,P(E/1) + b\,P(E/0)$$

Let the received photocurrent for a "1" state be represented by i_{p1}, and the RMS noise by i_{n1}. Similarly, define the photocurrent for a "0" state by i_{p0}, and the RMS noise by i_{n0}. It is often assumed that $i_{p0} = i_{n0}$.

The use of Gaussian noise distributions leads to the following form for the probability of error:

$$P(E) = \frac{1}{2}(1 - \text{erfc}(Q))$$

The complementary error function, erfc, is defined as:

$$\text{erfc}(z) = \frac{2}{\pi}\int_{z}^{\infty} e^{\frac{-x^2}{2}}\,dx$$

Tables of values for erfc(z) can be found in many books containing tables of mathematical functions.

The BER can be simplified for values lower than about 10^{-3} to the form:

$$P_e(Q) = \frac{1}{\sqrt{2\pi}}\frac{e^{\frac{-Q^2}{2}}}{Q}$$

The quality factor Q is defined as:

$$Q = \frac{1}{2}\sqrt{\mathrm{SNR}} = \frac{i_{po}}{2\,i_{no}} = \frac{i_{p1}}{2\,i_{n1}}$$

This simplified form is a close approximation to the full complementary error function form, and can be used for most designs. In particular, for a BER of 10^{-9}, the corresponding value of Q is 5.998.

D.1: Exercises

1. Write a simple computer program to calculate P_e given Q.
2. Make a plot of BER vs. Q for values of Q from 4 to 12.
3. What value of Q corresponds to a BER of 10^{-15}? What is the corresponding SNR?
4. A receiver has an SNR of 250. What is the corresponding Q and BER?
5. A receiver has an SNR of 30 dB. (a) What is the value of Q? (b) What is the BER?
6. A receiver has a BER of 10^{-12}. What are the corresponding Q and SNR?
7. Write a numerical methods program to calculate the SNR from the BER. (The equation for the BER cannot be solved directly for Q.)

Appendix E Maxwell's Equations

There are four equations describing the relationships between electric and magnetic fields. These equations were developed by James Clerk Maxwell. This was the first complete set of equations to gather all of the electric, magnetic, and electromagnetic "laws" (e. g. Ampere's law, the Biot–Savart law). In the differential form, given in Figure E-1, these equations apply at all points in space and at all points in time. There are four field variables (**D** and **E** for electric fields, **B** and **H** for magnetic fields. **D** and **E** are coupled through the relationship:

$$\mathbf{D} = \varepsilon_r \varepsilon_o \mathbf{E} = \varepsilon \mathbf{E}$$

Similarly, **B** and **H** are coupled through the relationship:

$$\mathbf{B} = \mu_r \mu_o \mathbf{H} = \mu \mathbf{H}$$

D, **E**, **B**, and **H** are all vector fields. This means that each field has x, y, and z components. In an *isotropic* medium, **D** is parallel to **E** and **B** is parallel to **H**, Maxwell's equations can be used to derive boundary conditions for electromagnetic fields. The fields do not maintain a constant value as they cross boundaries because a surface charge or surface current may exist at the boundary. For example, the reason **E** = 0 inside a metal is because there are many free electrons which move towards (or away from) the surface in such a way as to neutralize internal fields. Similarly, superconducting materials develop surface currents to exclude magnetic fields from their interiors. Figure E-2 shows the boundary conditions when there are no externally generated surface charges or currents.

$$\nabla \cdot \vec{D} = \rho$$

$$\nabla \cdot \vec{B} = 0$$

$$\nabla \times \vec{E} = -\frac{\partial \vec{B}}{\partial t}$$

$$\nabla \times \vec{H} = \frac{\partial \vec{E}}{\partial t} - \vec{J}_s$$

Figure E-1: Maxwell's Equations.

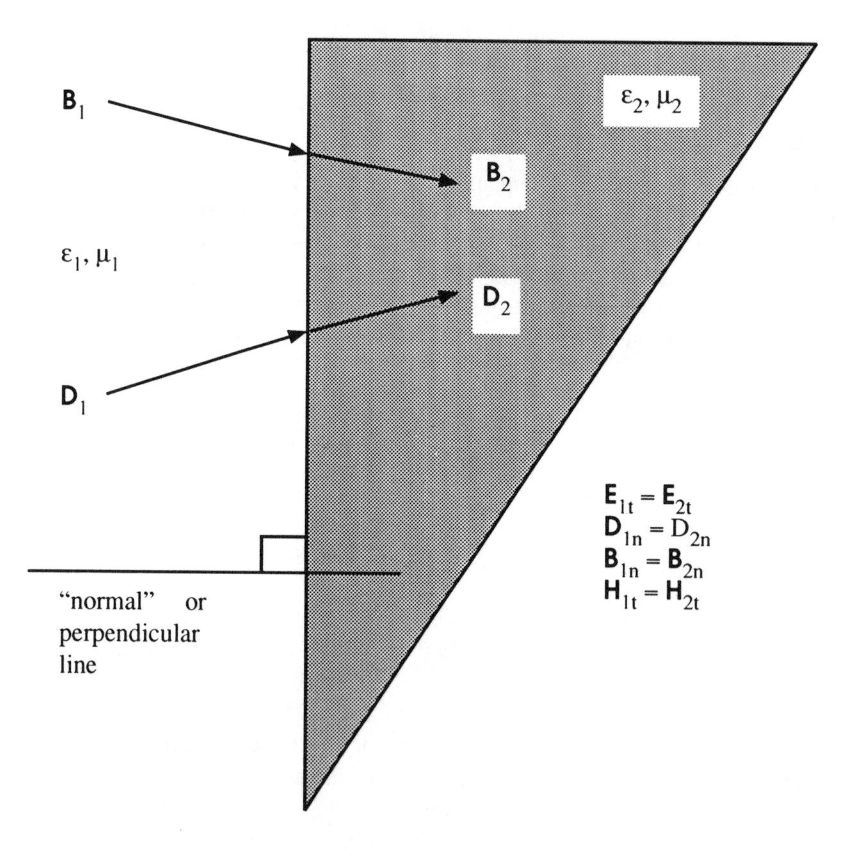

Figure E-2: Boundary Conditions.

When the permittivity (dielectric constant) changes across the dielectric boundary shown above, the tangential component of **E** is the same for each side of the boundary. $\mathbf{E}_t$ must remain the same because it is a direct measure of the force on a "test" particle at the boundary, and this force has the same value independent of how it is calculated. This implies that a change in dielectric always produces a surface charge density ρ at the dielectric boundary. Similarly, if the permeability changes across a boundary, a surface current must flow when a magnetic field is present, since $\mathbf{H}_n$ must be the same at the boundary for each region.

Three derivations using Maxwell's equations are presented here for propagating waves of the form $\mathbf{a}Ae^{j(\omega t-\mathbf{k}\cdot\mathbf{r})}$, where $\mathbf{a}$ is the field direction, A is the field amplitude, ω is the angular frequency ($2\pi f$), **k** is the direction of propagation, and **r** is the coordinate position. The first derivation shows that **E**, **B**, and **k** are *always* mutually perpendicular for a wave propagating in free space. The second shows that **E**, **B**, and **k** are *never* mutually perpendicular in a rectangular metal waveguide. The third derives the Bessel function solutions for a circular dielectric waveguide.

E.1: Maxwell's Equations: Free Space

Free space is characterized by $\varepsilon = \varepsilon_0$, $\mu = \mu_0$, $\rho = 0$, $\mathbf{J} = 0$. The general form of a propagating wave is $\mathbf{a}\,A\,e^{j(\omega t-\mathbf{k}\cdot\mathbf{r})}$. The electric and magnetic fields are characterized by direction of movement, **k**, the strengths E_0 and B_0, and directions of the fields, $\mathbf{a}_E$ and $\mathbf{a}_B$.

A homogeneous, isotropic, linear medium is similar to free space, except that the values of ε and μ are not the free space values. In the absence of charge and current, Maxwell's equations will have the same solutions as for free space, with the appropriate values for ε and μ.

Each of Maxwell's equations can be applied to these vectors in any coordinate space; for convenience, this derivation will be done in Cartesian coordinates. Showing that two vectors are perpendicular is done by showing that their dot product is identically zero or that one is the result of a cross–product operation on the other.

(1) $$\nabla\cdot \mathrm{D} = \nabla\cdot \epsilon \mathrm{E} = \nabla\cdot a_E E_o e^{j(\omega t-\mathrm{k}\cdot\mathrm{r})} = 0$$

$$= \left[\frac{\partial}{\partial \mathrm{x}}a_\mathrm{x} + \frac{\partial}{\partial \mathrm{y}}a_\mathrm{y} + \frac{\partial}{\partial \mathrm{z}}a_\mathrm{z}\right]\cdot a_E E_o e^{j\omega t} e^{j(\mathrm{k_x\,x + k_y\,y + k_z\,z})}$$

$$= E_o[\mathrm{k_x}\, a_{E\,x} + \mathrm{k_y}\, a_{E\,y} + \mathrm{k_z}\, a_{E\,z}] = E_o(\mathrm{k}\cdot a_E)$$

Clearly, E_0 is not zero, and therefore $\mathbf{k} \cdot \mathbf{a}_E = 0$, and $\mathbf{k}$ is perpendicular to $\mathbf{a}_E$. Similarly,

$$(2) \qquad \nabla \cdot B = \nabla \cdot a_B B_o e^{j(\omega t - k \cdot r)} = 0$$

$$= B_o[k_x a_{Bx} + k_y a_{By} + k_z a_{Bz}] = B_o(k \cdot a_B)$$

Again, B_0 is not zero, and therefore $\mathbf{k} \cdot \mathbf{a}_B = 0$, and $\mathbf{k}$ is perpendicular to $\mathbf{a}_B$.

Similarly:

$$(3) \qquad \nabla \times B = -\frac{\partial E}{\partial t} = j\omega E$$

$$\nabla \times B = a_x[k_y B_z - k_z B_y] + a_y[k_z B_x - k_x B_z] + a_z[k_x B_y - k_y B_x] = k \times B$$

Since $\mathbf{k} \times \mathbf{B} = j\omega \mathbf{E}$, $\mathbf{E}$ is perpendicular to both $\mathbf{k}$ and $\mathbf{B}$.

A wave equation can be derived for the time–dependent fields using Maxwell's equations and the vector identity:

$$\nabla \times \nabla \times E = \nabla(\nabla \cdot E) - \nabla^2 E$$

Taking the curl of both sides of Maxwells' 3rd equation gives:

$$\nabla \times (\nabla \times E) = -\nabla \times \left(\frac{\partial B}{\partial t}\right) = -\mu \frac{\partial}{\partial t}(\nabla \times H)$$

Substituting from Maxwells' 4th equation gives and noting that $\nabla \cdot E = 0$ gives:

$$\nabla^2 E = \epsilon\mu \frac{\partial^2 E}{\partial t^2}$$

Similarly, taking the curl of both sides of Maxwells' 4th equation and substituting from Maxwells' 3rd equation gives:

$$\nabla^2 H = \epsilon\mu \frac{\partial^2 H}{\partial t^2}$$

These two equations are known as the standard wave equations for homogeneous, isotropic, linear media.

E.2: Maxwell's Equations: Rectangular Metal Waveguide

A metal waveguide is characterized by the same properties as free space inside waveguide, and by perfect metal walls. At the wall boundaries, the electric field is exactly zero, since both the tangential and normal electric field is exactly zero in the metal walls. Inside the waveguide cavity, the fields obey the standard wave equations.

The rectangular waveguide has dimensions a and b in the x and y dimensions, respectively. The x axis is defined as the longer side (a > b). The coordinate system will be chosen with its origin at one corner of the waveguide, as shown in Figure E-3.

For a propagating wave, the amplitude varies as $e^{j\beta z}$. The propagation constant β is k_z inside the waveguide. (Note that k inside the waveguide is not the same as k outside the waveguide.) Finally, **E** and **B** are divided into component parts for ease of analysis:

$$\mathbf{E} = \mathbf{a}_E\, E_o\, e^{j(\omega t - \mathbf{k}\cdot\mathbf{r})} = [\, \mathbf{a}_x\, E_x\,(x,y,z) + \mathbf{a}_y\, E_y\,(x,y,z) + \mathbf{a}_z\, E_z\,(x,y,z)\,]\, e^{j\omega t}\, e^{j\beta z}$$

$$\mathbf{B} = \mathbf{a}_B\, B_o\, e^{j(\omega t - \mathbf{k}\cdot\mathbf{r})} = [\, \mathbf{a}_x\, B_x\,(x,y,z) + \mathbf{a}_y\, B_y\,(x,y,z) + \mathbf{a}_z\, B_z\,(x,y,z)\,]\, e^{j\omega t}\, e^{j\beta z}$$

Inside the waveguide, there is no charge density or current density. At the walls, there will be a surface charge and surface current density in the metal, as charges move to create a zero electric field at the metal surface. The electric field boundary conditions require the tangential electric field to be zero at all edges. the normal electric field is not required to be zero.

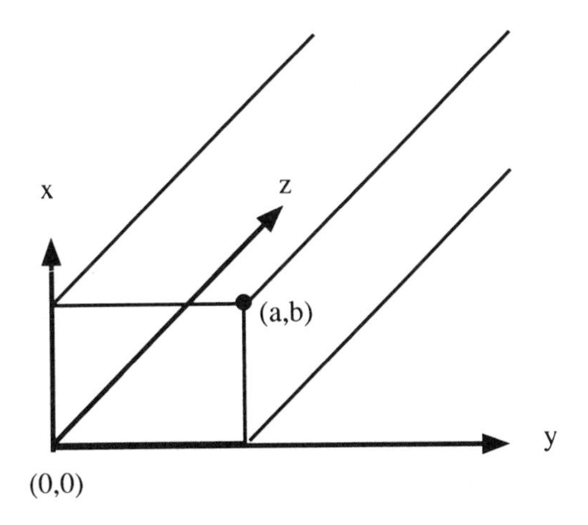

Figure E-3: Rectangular Waveguide Coordinate System.

The solutions are found by solving Maxwell's equations in rectangular coordinates and applying these boundary conditions. The solutions are always of the form:

$$\mathbf{E} = [\, \mathbf{a}_x E_x(x,y) + \mathbf{a}_y E_y(x,y) + \mathbf{a}_z E_z(x,y) \,]\, e^{j\omega t} e^{j\beta z}$$

$$\mathbf{B} = [\, \mathbf{a}_x B_x(x,y) + \mathbf{a}_y B_y(x,y) + \mathbf{a}_z B_z(x,y) \,]\, e^{j\omega t} e^{j\beta z}$$

There are three possible solutions: $\mathbf{E}_z = \mathbf{B}_z = 0$, $\mathbf{E}_z = 0$, or $\mathbf{B}_z = 0$. For the case $\mathbf{E}_z = 0$, the following equations are the solutions to Maxwell's equations in a rectangular waveguide:

$$E_x = E_{xo} \cos\left(\frac{m\pi x}{a}\right) \sin\left(\frac{n\pi y}{b}\right)$$

$$E_y = E_{yo} \sin\left(\frac{m\pi x}{a}\right) \cos\left(\frac{n\pi y}{b}\right)$$

$$B_x = B_{xo} \sin\left(\frac{m\pi x}{a}\right) \cos\left(\frac{n\pi y}{b}\right)$$

$$B_y = B_{yo} \cos\left(\frac{m\pi x}{a}\right) \sin\left(\frac{n\pi y}{b}\right)$$

The z components of the fields are given by:

$$E_z = 0$$

$$B_z = B_{zo} \cos\left(\frac{m\pi x}{a}\right) \sin\left(\frac{n\pi y}{b}\right)$$

The solutions for the case $\mathbf{B}_z = 0$ are similar in form, with a and b swapped in the equations. For both TE and TM modes, the solutions require m and n to be positive integers or zero. The propagation constant in the waveguide is given by:

$$\beta = \sqrt{k_o^2 - \left(\frac{m\pi}{a}\right)^2 - \left(\frac{n\pi}{b}\right)^2} \qquad k_o = 2\pi f\sqrt{\epsilon\mu}$$

where k_o is the propagation constant for a plane wave in free space. Since either m or n is non–zero, the waveguide propagation constant is smaller than the free space propagation constant. This also implies a cutoff condition for β and ω, which is $\beta_c = 0$ and:

$$f_c = \frac{1}{2\sqrt{\epsilon\mu}} \sqrt{\left(\frac{m}{a}\right)^2 + \left(\frac{n}{b}\right)^2}$$

The solutions to Maxwell's equations in the rectangular guide are either transverse electric (E_z=0) or transverse magnetic (B_z=0) waves. Each solution is a mode of the waveguide. The lowest order mode, having the lowest cutoff frequency, is TE_{10} , which is a transverse electric field having m=1 and n=0. Each mode is characterized by the notation TE_{mn} or TM_{mn}.

For frequencies below the TE_{10} cutoff frequency, wave propagation occurs with very high attenuation. For frequencies above 2 TE_{10}, there will always be more than one permitted mode. In most metal waveguides, the waveguide is designed so that the cutoff frequency for the TE_{01} mode is equal to twice that of the TE_{10}; thus the waveguide carries a single mode for frequencies between 1 and 2 times f_c. When more than one mode is present, the link bandwidth is severely limited by modal dispersion effects.

E.3: Maxwell's Equations: Circular Dielectric Waveguide

A dielectric waveguide is characterized by the dielectric constant inside the waveguide, and by a concentric wall having a lower dielectric constant. At the wall boundaries, the electric field is not zero, but the tangential E and normal D fields must match at the dielectric boundary.

The circular waveguide has radius a. Since the waveguide has a cylindrical shape, a cylindrical coordinate system will be used. The coordinate system will be chosen with its origin at the center of the waveguide, as shown in Figure E-4.

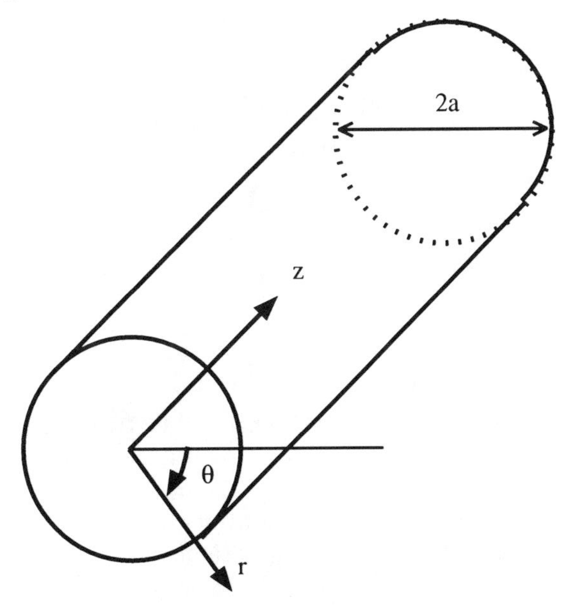

Figure E-4: Cylindrical Waveguide Coordinate System.

Inside the waveguide, there is no charge density or current density. At the walls, there will be a surface charge but no surface current density in the dielectric boundary. The resulting electric field boundary conditions are:

$$E_{\theta 1}\ (a,\theta,z) = E_{\theta 2}\ (a,\theta,z)$$

$$E_{z1}\ (a,\theta,z) = E_{z2}\ (a,\theta,z)$$

$$\varepsilon_1\ E_{r1}\ (a,\theta,z) = \varepsilon_2\ E_{r2}\ (a,\theta,z)$$

The solutions are found by solving Maxwell's equations in cylindrical coordinates and applying these boundary conditions. The solutions are always of the form:

$$\mathbf{E} = [\ \mathbf{a}_r\ E_r(r,\theta) + \mathbf{a}_\theta\ E_\theta\ (r,\theta) + \mathbf{a}_z\ E_z(r,\theta)\ ]\ e^{j\omega t}\ e^{j\beta z}$$

$$\mathbf{B} = [\ \mathbf{a}_r\ B_r(r,\theta) + \mathbf{a}_\theta\ B_\theta\ (r,\theta) + \mathbf{a}_z\ B_z(r,\theta)\ ]\ e^{j\omega t}\ e^{j\beta z}$$

These equations are substituted into Maxwell's equations, and it is found that the z components of the fields must satisfy the following conditions:

$$\frac{\partial^2 E_z}{\partial r^2} + \frac{1}{r}\frac{\partial E_z}{\partial r} + \frac{1}{r^2}\frac{\partial^2 E_z}{\partial \phi^2} + q^2\, E_z = 0$$

$$\frac{\partial^2 H_z}{\partial r^2} + \frac{1}{r}\frac{\partial H_z}{\partial r} + \frac{1}{r^2}\frac{\partial^2 H_z}{\partial \phi^2} + q^2\, H_z = 0$$

$$q = (2\pi f)^2 \epsilon \mu - \beta^2$$

The solutions to these equations consist of the Bessel functions J_v in the core and K_v in the cladding. J_v has an oscillatory nature, with amplitude decreasing with increasing radius. K_v does not oscillate, and has an amplitude decreasing with increasing radius outward from the core. J_v is similar to an exponentially damped sinusoid, while K_v is similar to a decaying exponential. For the lowest order Bessel function, J_o, the value at the center of the core $J_o(0) = 1$, while for all other orders $J_v(0) = 0$.

The following equations are the solutions to Maxwell's equations under these constraints:

$$E_z = A J_\nu(ur)\, e^{j\nu\phi}\, e^{j(\omega t - \beta z)} \qquad r < a$$

$$B_z = C J_\nu(ur)\, e^{j\nu\phi}\, e^{j(\omega t - \beta z)} \qquad r < a$$

$$E_z = F K_\nu(wr)\, e^{j\nu\phi}\, e^{j(\omega t - \beta z)} \qquad r > a$$

$$B_z = G K_\nu(wr)\, e^{j\nu\phi}\, e^{j(\omega t - \beta z)} \qquad r > a$$

Neither $E_z = 0$ nor $B_z = 0$. The solutions require ν to be a positive integer or zero. The permitted values of β are given by:

$$n_2 \frac{2\pi}{\lambda} \leq \beta \leq n_1 \frac{2\pi}{\lambda}$$

The scale factors u and w in the Bessel function arguments are related to the propagation constant in the waveguide:

$$u = \sqrt{\left(\frac{n_1 2\pi}{\lambda}\right)^2 - \beta^2}$$

$$w = \sqrt{\beta^2 - \left(\frac{n_2 2\pi}{\lambda}\right)^2}$$

The boundary conditions at the dielectric interface lead to constraints on u, w, β, and on the relative field strengths A, C, F, and G.

The cutoff condition for single–mode operation corresponds to the first zero of $J_0(ua)$, where V = 2.405. Due to the dielectric boundaries, energy propagates without loss for frequencies both above and below cutoff . Below the cutoff frequency (above the cutoff wavelength), only one mode propagates. Above the cutoff frequency, more than one mode can propagate.

E.4: Exercises

1. Show that $\varepsilon_0 \mu_0 = 1/c^2$.
2. Write the boundary conditions for the magnetic field at the walls of a rectangular metal waveguide.
3. Find the cutoff frequency for a metal waveguide of dimensions 2 x 5 cm.

4. Find the surface charge density at time = 0 and position = (0,1,0) for a metal waveguide of dimensions a = 5 cm and b = 2 cm, carrying a TE_{10} wave at 12 GHz.
5. Write the boundary conditions for the magnetic field at rectangular metal waveguide walls.
6. Write the boundary conditions for the magnetic field at cylindrical dielectric waveguide walls.
7. Look up the values of the Bessel functions J_v (x), and plot the first three Bessel functions from x = 0 to x = 5.
8. Look up the values of the Bessel functions K_v (x), and plot the first three Bessel functions from x = 0 to x = 3.
9. Find the cutoff frequency and wavelength for a fiber having a 6 μm diameter, a core index of 1.447, and a cladding index of 1.441.

Appendix F Frequency Response

The frequency response of an analog system can be expressed in many ways. These ways include characteristic time constants and corner frequencies, plots of the transfer function vs. frequency, pole–zero plots, and transfer functions (often written in factored form). All of these ways of expressing the frequency response assume a linear system (output proportional to input).

A simple RC circuit, such as that shown in Figure F-1, has a frequency response similar to that of many fiber optic links. The circuit response has a single RC time constant ($\tau = RC$). If a step input is applied at V_{in}, then the output voltage will respond with the classic rising RC curve seen in circuit analysis books.

The simple RC circuit can also be characterized by its pole frequency:

$$f_{p1} = \frac{1}{2\pi RC}$$

If an AC (sinusoidal) input is applied at V_{in}, then the output voltage amplitude will be at $0.707V_{in}$ when the input frequency is f_{p1}. This voltage is 3 dB down from the DC value.

For a system with a single time constant, the frequency f_{p1} is referred to as the pole frequency or the 3–dB frequency. In a system having more than one pole, the output will be 3 dB down at a frequency that is lower than the first pole frequency. The 3–dB frequency is also referred to as the corner frequency.

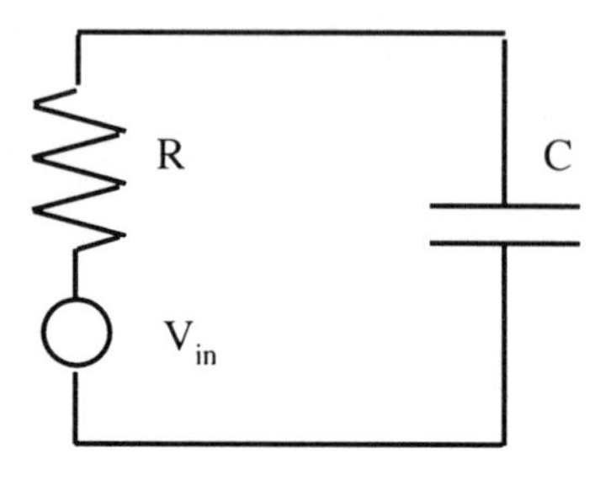

Figure F-1: Simple RC Circuit.

A plot of output vs. frequency for the simple RC circuit is shown in Figure F-2. When the amplitude is plotted in dB and the frequency is plotted on a logarithmic scale, the frequency plot can be approximated by two straight lines, as shown in the figure. The slope of the line below the pole is 0 dB/decade, while the slope above the pole is 20 dB/decade down.

The (steady state) frequency response for the simple RC circuit is given by:

$$H(f) = \frac{V_{out}}{V_{in}} = \frac{1}{1 + \dfrac{jf}{2\pi RC}}$$

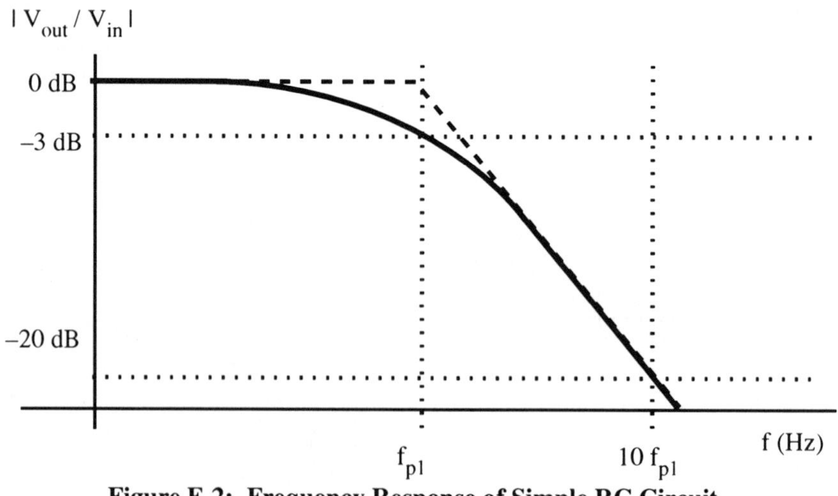

Figure F-2: Frequency Response of Simple RC Circuit.

Appendix G Semiconductor Materials

Semiconductor materials can be classified in a number of ways. There are two major classifications of interest in fiber optics. The first is the division into direct and indirect bandgap materials. The second is into elemental and compound semiconductors.

The energy and momentum of carriers (holes and electrons) in a semiconductor are related through bandgap diagrams such as those shown in Figure G-1. Electrons are found at the bottom of the lowest valley of the conduction band, and holes at the top of the valence band. When the bottom of the conduction band is centered over the top of the valence band, the semiconductor has a direct bandgap. When an electron and hole combine to form a chemical bond, there is almost no momentum difference. Since a photon carries only a small amount of momentum, it can be given off during a direct recombination.

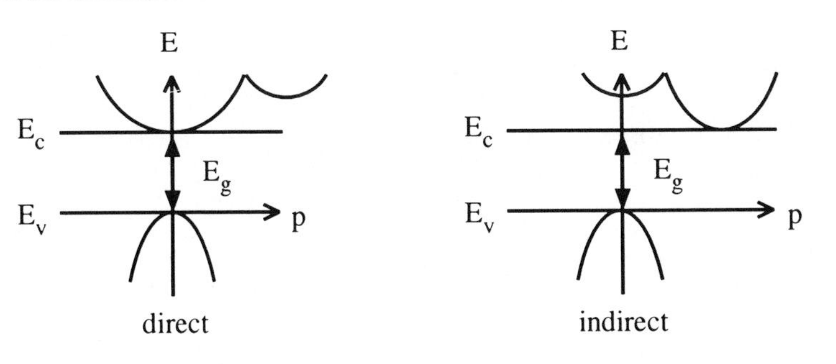

Figure G-1: Energy vs. Momentum.

In an indirect bandgap material, the lowest valley of the conduction band is displaced from the top of the valance band. In this case, the recombination event requires the removal of a larger momentum than a photon can carry off. The interaction is still a three–body problem, but the released energy and momentum appear as heat (phonons).

The active (light emitting) regions of LEDs and lasers must be made from direct bandgap semiconductor materials. Indirect bandgap materials may be used in other layers in a heterojunction device. Photodiodes, on the other hand, may be made from either type of material, and are most commonly made from indirect materials to maximize responsivity.

The second classification of semiconductors involves the number of chemical elements used to make the crystal. In an elemental semiconductor, only a single chemical element is used. Most elemental semiconductors are made from silicon or germanium, since the bandgaps are in the region of interest for fiber optic communications.

Compound semiconductors are more expensive to make, since the mixture of elements cannot be exactly controlled in most cases. The easiest to control is a binary (two element) semiconductor, such as GaAs. Binary compounds require a precise balance of content from the column III element and the column V element. Binary compounds can also be made from column II and column VI compounds.

Ternary and quaternary compounds are much more difficult to produce. Not only is it necessary to balance the column III elements and column V elements, but it is also necessary to balance the proportions between the column III elements and between the column V elements individually. Only certain chemical ratios of the elements will create a direct bandgap material; a change in proportions of the same chemicals can affect both the size and directness of the bandgap.

Bandgap engineering involves the selection of the appropriate elements and ratios to achieve specified material properties. The properties may include the bandgap, the directness, the index of refraction, the thermal expansion coefficient, or other properties which may influence performance or reliability. Table G-1 gives some of the properties of the more commonly used semiconductors.

Table G-1: Properties of Semiconductors

Material	Bandgap energy	Bandgap type	ε_r
Si	1.12 eV	indirect	11.7
Ge	0.67 eV	indirect	16.3
Ga As	1.43 eV	direct	13.2
Ga Sb	0.7	direct	15.7
In P	1.28	direct	12.4
In As	0.36	direct	14.6
Cd Te	1.58	direct	10.2
$In_{0.74}\,Ga_{0.26}\,As_{0.56}\,P_{0.44}$	0.96 eV	direct	___
Si C	2.8	direct	10.0
$Al_{1-x}\,Ga_x\,P$ ($x < 0.36$)	1.42–1.87 eV	direct	___
$Al_{1-x}\,Ga_x\,P$ ($x > 0.36$)	___	indirect	___

Appendix H Power and Risetime Budget Sheets (Blanks)

System Specifications

Link distance	____	km
Data Rate	____	MHz or Mbit/sec
Coding	____	
Required system risetime	____	nsec
Required system quality	____	BER or SNR
System design margin	____	dB

System Costs

Transmitter	$ ________
Receiver	$ ________
Connectors & splices	$ ________
Fiber	$ ________
Electronics	$ ________
Installation & maintenance	$ ________

Power and Risetime Budget Sheets (Blanks)

Power (Loss) Budget

Transmitter power	____	mW	____	dBm
Fiber coupling (NA effect)	____	dB		
Connector losses			____	dB
____ # connectors				
____ Loss / connector				
Fiber attenuation			____	dB
____ length (km)				
____ dB / km				
Splice losses			____	dB
____ # splices				
____ Loss / splice				
Other losses (e. g. bending, splitters, etc.)			____	dB
Gains (e. g. fiber amplifier)			— ____	dB
Design margin			____	dB

System minimum output power	____	mW	____	dBm
Detector minimum acceptable output	____	dBm		
Excess margin			____	dB

Minimum power > required power? **Y / N**

Risetime (Bandwidth) Budget

Transmitter risetime			____	nsec
Receiver risetime			____	nsec
Fiber dispersion			____	nsec
Intermodal (0.44/B_0)	____	nsec / km		
Intramodal	____	nsec / km		
Waveguide	____	nsec / km		

System maximum risetime	____	nsec
Required risetime	____	nsec

Maximum risetime < required risetime? **Y / N**

INDEX